OFFICE INFORMATION SYSTEMS

Management Issues and Methods

Richard H. Irving
York University, Ontario

and

Christopher A. Higgins
University of Western Ontario

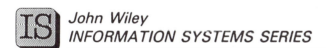

John Wiley
INFORMATION SYSTEMS SERIES

JOHN WILEY & SONS

Chichester · New York · Brisbane · Toronto · Singapore

Other Wiley Editorial Offices

John Wiley & Sons, Inc., 605 Third Avenue,
New York, NY 10158-0012, USA

Jacaranda Wiley Ltd, G.P.O. Box 859, Brisbane,
Queensland 4001, Australia

John Wiley & Sons (Canada) Ltd, 22 Worcester Road,
Rexdale, Ontario M9W 1L1, Canada

John Wiley & Sons (SEA) Pte Ltd, 37 Jalan Pemimpin #05-04,
Block B, Union Industrial Building, Singapore 2057

Library of Congress Cataloging-in-Publication Data:

Irving, R.
    Office information systems : Management issues and
methods / R. Irving, C. Higgins.
        p.   cm. — (John Wiley information systems series)
    Includes bibliographical references and index.
    ISBN 0-471-91888-1
    1. Business—Data processing.   2. Office practice—Automation.
I. Higgins, C.   II. Title.   III. Series.
HF5548.2.I7   1991
651.8—dc20   90–22243
                                                          CIP

British Library Cataloguing in Publication Data:

Irving, R.
    Office information systems : management issues and methods.
    1. Office practices. Information systems
    I. Title   II. Higgins, C.   III. Series
    651

    ISBN 0-471-91888-1

Printed and bound in Great Britain by Courier International Ltd, Tiptree, Essex

OFFICE INFORMATION SYSTEMS

John Wiley
INFORMATION SYSTEMS SERIES

Editors

Richard Boland
Case Western Reserve University

Rudy Hirschheim
University of Houston

To Pat and Barb

CONTENTS

SERIES FOREWORD

In order for all types of organisations to succeed, they need to be able to process data and use information effectively. This has become especially true in today's rapidly changing environment. In conducting their day-to-day operations, organisations use information for functions such as planning, controlling, organising, and decision making. Information, therefore, is unquestionably a critical resource in the operation of all organisations. Any means, mechanical or otherwise, which can help organisations process and manage information presents an opportunity they can ill afford to ignore.

The arrival of the computer and its use in data processing has been one of the most important organisational innovations in the past thirty years. The advent of computer-based data processing and information systems has led to organisations being able to cope with the vast quantities of information which they need to process and manage to survive. The field which has emerged to study this development is *information systems* (IS). It is a combination of two primary fields, computer science and management, with a host of supporting disciplines, e.g. psychology, sociology, statistics, political science, economics, philosophy, and mathematics. IS is concerned not only with the development of new information technologies but also with questions such as: how they can best be applied, how they should be managed, and what their wider implications are.

Partly because of the dynamic world in which we live (and the concomitant need to process more information), and partly because of the dramatic recent developments in information technology, e.g. personal computers, fourth-generation languages, relational databases, knowledge-based systems, and office automation, the relevance and importance of the field of information systems, and office automation, the relevance and importance of the field of information systems has become apparent. End users, who previously had little potential of becoming seriously involved and knowledgeable in information technology and systems, are now much more aware of and interested in the new technology. Individuals working in today's and tomorrow's organisations will be expected to have some understanding of and the ability to use the rapidly developing information technologies and systems. The dramatic increase in the availability and use of information technology, however, raises fundamental questions on the guiding of technological innovation, measuring organisational and managerial productivity, augmenting human intelligence, ensuring data integrity, and establishing strategic advantage. The expanded use of information systems also raises major challenges to the traditional forms of administration and authority, the right to privacy, the nature and form of work, and the limits of calculative rationality in modern organisations and society.

The Wiley Series on Information Systems has emerged to address these questions and challenges. It hopes to stimulate thought and discussion on the key role information systems play in the functioning of organisations and society, and how their role is likely to change in the future. This historical or evolutionary theme of the Series is important because considerable insight can be gained by attempting to understand the past. The Series will attempt to integrate both description—what has been done—with prescription—how best to develop and implement information systems.

The descriptive and historical aspect is considered vital because information systems of the past have not necessarily met with the success that was envisaged. Numerous writers postulate that a high proportion of systems are failures in one sense or another. Given their high cost of development and their importance to the day-to-day running of organisations, this situation must surely be unacceptable. Research into IS failure has concluded that the primary cause of failure is the lack of consideration given to the social and behavioral dimensions of IS. Far too much emphasis has been placed on their technical side. The result has been something of a shift in emphasis from a strictly technical conception of IS to one where it is recognised that information systems have behavioral consequences. But even this misses the mark. A growing number of researchers suggest that information systems are more appropriately conceived as social systems which rely, to a greater and greater extent, on new technology for their operation. It is this social orientation which is lacking in much of what is written about IS. The current volume, *Office Information Systems: Management Issues and Methods*, offers a practical view on office information systems. The authors provide a sensible and thoughtful treatment of how office information systems could be developed and used in organisations.

The Series seeks to promote a forum for the serious discussion of IS. Although the primary perspective is a more social and behavioral one, alternative perspectives will also be included. This is based on the belief that no one perspective can be totally complete; added insight is possible through the adoption of multiple views. Relevant areas to be addressed in the Series include (but are not limited to): the theoretical development of information systems, their practical application, the foundations and evolution of information systems, and IS innovation. Subjects such as systems design, systems analysis methodologies, information systems planning and management, office automation, project management, decision support systems, end-user computing, and informtion systems and society are key concerns of the Series.

Rudy Hirschheim
Richard Boland

PREFACE

This book grew from the authors' experiences with office technology over the past 15 years. Since the mid-1970s, starting with the development and proliferation of electronic mail systems, office information systems (as distinct from management information systems [MIS], data processing [DP] or electronic data processing [EDP]) have become a multi-billion-dollar industry. While the technology has progressed rapidly, our capability to use it wisely has progressed more slowly.

In our research, consulting and teaching we have found that both managers and students have problems coming to grips with the fundamental issues surrounding office information systems. The main problem is that the technology itself is extremely attractive. Learning a new system can be fun, and many people value the respect given their expertise once they master a system. Furthermore, aided by sophisticated marketing promotions, Office Information Systems have acquired a patina of progress, youth, and fast-track advancement. It is not too surprising that some managers buy into OIS as much for the image it creates as for its contribution to the bottom line.

There are signs that this heady atmosphere is cooling. Sales of personal computers are no longer growing at the phenomenal rate of a few years ago. Increasing familiarity with the technology, and several well-publicized disasters, have led to a reassessment of its merits. Given this cooler environment, the time is appropriate for a careful look at the management issues associated with OIS.

ACKNOWLEDGMENTS

Many people have contributed to this book. Foremost among these are Rudy Hirschheim and Dick Boland, with whom we first discussed the concept of a book on management issues arising from OIS. They read several drafts and have always had useful insights, suggestions and criticisms. We must also thank the students in MGTS6730 at York University who have, over the past few years, read and commented on various rough drafts. Their insights have materially improved the content. Catherine Snedden reviewed the manuscript in detail, made a number of important changes and suggested useful additions. At Wiley, Diane Taylor has proven consistently patient and helpful.

Others have had an indirect influence. Professor Dave Conrath at the University of Waterloo provided both authors with their first taste of office systems research. Professor Martin Elton at New York University inspired a rethinking of several fundamental assumptions about OIS. Professor Frank Safayeni at Waterloo provided inspiration by consistently taking an insightful and thoughtful approach to OIS. We must also thank the Canadian Federal Department of Communication, the Canadian Workplace Automation Research Centre, and the Social Sciences and Humanities Research Council for research funding they have provided over the years.

Finally, we thank our wives Pat and Barb, who make what we do worth while, and our children, who make it necessary.

Chapter 1
INTRODUCTION TO OIS

The growth of office systems technology presents managers with a complex variety of alternatives. Not only are the technical alternatives bewildering, the organizational implications are profound.

Many executives and managers are using the advent of office information systems (OIS) as an opportunity to reconsider their business operations. Unfortunately, other executives see technology as a panacea for various organizational ills. This is not the case. In fact, computerizing existing office systems may exacerbate organizational problems.

Strassmann (1985, p. 159) contends that 'there are already good reasons to conclude that there is no direct and simple correlation between management productivity and information technology.' He suggests that office information systems should be installed only after the business has been improved by other means. This position is also echoed by Baetz (1985, p. 8) who writes: 'Whether your people, and you, can do what is required more easily, more competently, and more flexibly with the tools you have chosen is the only question that should concern you.'

Introducing an OIS without a clear understanding of the business is a recipe for disaster. The most important question one can ask is:

HOW CAN WE IMPROVE THE OVERALL FUNCTIONING OF OUR ORGANIZATION?

Only when this question has been addressed should one deal with the secondary issue:

CAN OIS TECHNOLOGY FACILITATE THE IMPROVEMENT?

In other words, the goals and aims of the organization should drive the choice (if any) of OIS. Clearly, technology in search of a problem is not the answer. Solutions lie in the vision, commitment and fundamental business skills of senior executives. When these are coupled with an informed and involved workforce, the potential for substantial gains exist. This perspective requires that we turn our attention from the technology and address the management issues raised by OIS.

The issues that managers must face when considering office information systems are dealt with in the chapters which follow. We identify the potholes in the road and suggest some means of avoiding them; however, each reader must formulate a strategy for his or her own organization.

ORGANIZATION OF THIS BOOK

The book consists of four parts. Part I addresses the fundamental issues surrounding office information systems. Four chapters deal with issues that are fundamental to understanding OIS and the context of their use. These are: 'The Organizational Context of OIS', 'Office Technology and the Integration of Business Systems', 'Strategic Use of OIS', and 'Consequences of OIS.' This part provides the 'Why' of OIS.

Part II describes the operational issues with which managers must deal to successfully plan, implement and operate an OIS. Three chapters discuss 'Planning for OIS', 'Needs Analysis for OIS: A Management Perspective', and 'Implementing and Evaluating OIS.' This part provides instruction in the 'How to' aspects of OIS.

Part III presents three chapters dealing with developing issues. Each chapter, 'Using Computers to Monitor Employee Performance', 'Flexible Work and OIS' and 'End-user Computing', presents management issues which are beginning to emerge as major forces in the development of OIS. Part IV contains one chapter, 'Management and OIS: Where do We Go From Here?' In this chapter the material from the previous 11 chapters is summarized briefly and the management implications of future trends are discussed.

The chapters have been arranged in a sequence that makes sense to the authors. Individual readers may wish to read them in a different order, or to skip some chapters and concentrate on others. To facilitate this process a brief synopsis of each chapter follows.

Chapter 2 provides a grounding in organizational issues that surround OIS and provides a mechanism (the feasibility triangle) so that the non-technical reader can identify the feasible region for OIS. It also presents a process (based on organizational parameters) for assessing the potential scope for OIS. These two ballpark assessments provide a managerial basis for controlling the planning and needs analysis processes discussed in Chapters 6 and 7.

Chapter 3 provides a conceptual overview of OIS technology and issues arising from the potential integration of business systems. The focus is on explaining technology in a generic manner to readers who lack hands-on experience. Readers familiar with computer technology may wish to skip this section. The last section of the chapter addresses systems integration from a business perspective. The question of the balance between automation and augmentation is raised, as are generic technical issues.

Chapter 4 focuses on the strategic use of OIS technology. It begins by reviewing the use of technology for competitive advantage. The focus then turns to the strategic use of information technology. On the basis of this discussion the strategic implications of OIS technology are presented. The authors argue that OIS is an enabling technology which has indirect strategic potential for most organizations.

Chapter 5 presents and discusses the societal and organizational consequences of OIS. The aim of the discussion is to sensitize managers to the possible consequences of technical decisions and to provide a set of issues which will focus attention on key social, political and organizational problems.

Chapter 6 outlines a methodology for planning OIS. It covers all stages from preliminary assessment to implementation. The focus here is prescriptive in the sense that specific guidelines are presented. The underlying philosophy is that OIS planning can occur only with the active support of senior management and the meaningful involvement of all levels of the organizational hierarchy.

Chapter 7 describes the process of needs analysis and presents a methodology for conducting an analysis at the workgroup level. The main value of this approach is that data on key tasks and communication links are collected through a process of social consensus. These data provide the basis for senior management to conduct a critical organizational analysis and set priorities for OIS development. Information is also provided at the operational level which assists the IS group in targeting resources for detailed systems planning and design activities. The process is illustrated by means of a detailed case history.

Chapter 8 discusses the problems and issues involved in implementing and evaluating OIS. It expands on the material presented in Chapters 6 and 7. Key elements discussed in the implementation process are management of the change process, development of education and training programs, and management of the physical implementation process. Key elements discussed in the evaluation process are definition of performance criteria, data collection and analysis, and managerial assessment of the data.

Chapter 9 presents the developing issue of computerized performance monitoring. This chapter shows how computerized performance monitoring is growing, shows how it affects all levels in the organization, and discusses the positive and negative consequences. The underlying assumption is that computerized monitoring, if used wisely, can have positive effects both for employees and managers. The authors are careful to point to the likely negative results if computerized monitoring is not used appropriately. Rules for designing and implementing these systems are presented.

Chapter 10 contains a discussion of flexible work. The focus is on various forms of telecommuting and the management issues associated with them. Current literature is reviewed and discussed in terms of relevance to management. The theme is on the necessity to balance the benefits to the organization and benefits to employees if a telecommuting program is to be successful.

Chapter 11 covers the issue of end-user computing and the management issues that it raises. Advantages and disadvantages are discussed. Subsequently, strategies for managing this phenomenon are presented and discussed.

Chapter 12 summarizes the concepts presented in earlier chapters and presents a view of the future of OIS. This view is based on the assumption that more fully integrated systems will speed up the decision-making process and at the same time make it more complex. Consequently, management of decision processes will absorb an increasing proportion of managerial attention as the century comes to a close.

The Appendix presents an historical perspective on performance monitoring and provides background for readers who wish to trace the development of this issue.

BIBLIOGRAPHY

Baetz, M. L. (1985) *The Human Imperative: Planning for People in the Electronic Office*, Holt, Rinehart and Winston, Toronto, Canada.
Strassmann, P. A. (1985) *Information Payoff*, Free Press, New York.

Part I
FUNDAMENTAL ISSUES IN OIS

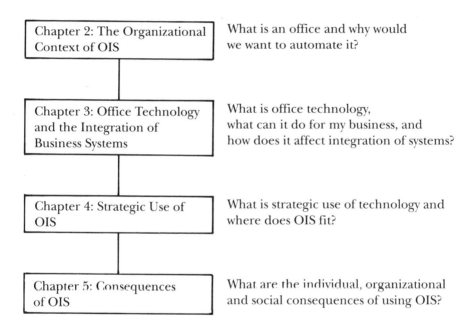

Chapter 2: The Organizational Context of OIS — What is an office and why would we want to automate it?

Chapter 3: Office Technology and the Integration of Business Systems — What is office technology, what can it do for my business, and how does it affect integration of systems?

Chapter 4: Strategic Use of OIS — What is strategic use of technology and where does OIS fit?

Chapter 5: Consequences of OIS — What are the individual, organizational and social consequences of using OIS?

Chapter 2
THE ORGANIZATIONAL CONTEXT OF OIS

A basic understanding of office information systems (OIS) and the organizational context in which they are used is a prerequisite for their successful design and implementation. This chapter begins by answering the question: What is an office information system?' We then describe an organizational scan which identifies the parameters constraining the development of office information systems. Following the organizational scan, a discussion of the factors to consider in defining the preliminary scope for OIS projects ensues. The chapter concludes with a summary of key management issues and an illustrative case history.

WHAT IS AN OFFICE INFORMATION SYSTEM?

In an office, information moves from person to person, forms are completed, filed, and retrieved, and individuals communicate a variety of both simple and complex concepts. At the turn of the century most written information was compiled by hand in pen and ink, copies were made by hand, and communication was by letter, telegraph or in person.

Today, many of the tedious functions related to information handling and processing are done through computer and telecommunication systems. Communications systems have evolved to the point where a manager can be in touch with his business from his or her home, car, or in midflight. However, the central issues underlying managerial activities are the same as they have always been; to make the business function better.

Computers and telecommunications systems have made it possible for large and complex organizations to control their operations and handle large volumes of business with relative ease. The tendency of writers to focus on the technology which supports business systems has generated such terms as 'office automation', 'the office of the future' and 'integrated office systems.' It is useful to note that the most integrated office system consists of the one person manual office where the boss carries everything in his or her head.

Definition of OIS

When we refer to OIS we refer to those information and communication systems which are necessary for the functioning of the office. We assume that telecommunications, data processing, and word processing play a large role; however we recognize that *human* systems underlie these *technical* systems.

In an ideal situation the technology and the people function together harmoniously. This seamless integration of technical systems and human systems is what *we* mean by an OIS. Its elements include relational databases, personal computing, electronic mail, telecommunications systems, FAX, and appropriate organizational and interpersonal structures. For these elements to function together smoothly, such technical issues as voice–data integration, standardization of software, and compatibility of hardware and software must be addressed. Human issues such as the design of individual jobs, the evolution of new organizational structures and the development of a new strategic vision of the organization must be dealt with. These considerations lead us to the following definition:

> *An Office Information System is a seamless integration of telecommunications, data processing, and personal computing with manual business processes; which supports key business functions; and which improves effectiveness, efficiency, and quality of working life.*

Implicit in our definition are the assumptions that employees are involved in the design of the system, and that computer-based systems are *integrated* with manual systems. It is crucial that a successful OIS should enhance the quality of work life for a majority of employees while facilitating business processes. One key element in achieving this goal is identifying the appropriate balance between automating and augmenting human work.

The Balance Between Augmentation and Automation

The choice between automating activities and augmenting human effort reflects a philosophical consideration which is often overlooked and which is related to the general management philosophy of the firm.

Automating replaces human effort with technology while augmentation facilitates human effort with technology by making work easier, faster, or perhaps even possible. The choice of one over the other is more than a technical decision. It strikes to the core of what you believe an office to be. Some managers and executives see an office as a factory where technology should be used to replace human effort wherever possible. Others view the office as a complex human environment where a constant balance must be maintained between the demands of the technology and the needs of humans.

In principle one automates routine, mindless work and augments complex sophisticated activity. In practice the choice is not so simple. The complexities of the problem are illustrated by attempts at automating routine clerical work such as processing insurance claims. In organizations where the automation model was followed closely, workers' jobs were substantially degraded and a negative work environment was generated. In one organization we visited, workers' washroom breaks were timed and no conversations were allowed during work periods. The rationale was that conversations could distract some workers from attention to their terminals. Automation, used in this fashion, creates new 'sweatshops', replicating the impersonal atmosphere common in the industrial factories in the early part of this century.

This need not have been the case. In other organizations with similar work, clerical workers find their jobs enhanced. Time saved from routine efforts is used to improve customer relations and service. In these latter cases the computer systems are similar to those of the former organizations; the difference lies in the way these systems and the employees who use them are managed (see Chapter 9 for a detailed discussion of these issues).

The Organizational Context

Understanding the organizational context of OIS is a prerequisite to its successful design, implementation and adoption. Hirschheim (1985, ch. 3) discusses a number of views of the office and suggests that an 'interpretivist' perspective provides a balance between the analytic and sociotechnical viewpoints. He subsequently (1985, ch. 4) reviews a number of office automation methodologies and models.

While we recommend Hirschheim's book as further reading, our purpose is not to review the various models of offices and methodologies for needs analysis. Here we present prescriptive approaches to organizational assessment, needs analysis and planning. In the remainder of this chapter we discuss the assessment approach: an organizational scan, and the definition of potential scope for OIS and show, by means of a case history, how these two activities are used by senior management. The other approaches are discussed in Chapter 6, which covers the detailed planning methodology for OIS, and in Chapter 7, which covers the detailed needs analysis methodology.

THE ORGANIZATIONAL SCAN AND POTENTIAL SCOPE OF OIS

The issues associated with OIS are complex. One reason for this is that the design, implementation and effective use of these systems is dependent on the level of resources available to support them, and the requirements of the particular organization in which they are implemented. In order to simplify the process we recommend that management do some preliminary homework. This consists of two parts: an organizational scan to determine ballpark figures on resource availability, and a preliminary definition of the potential scope of OIS to determine the likely demand for support. Each is discussed below.

The *organizational scan* defines the constraints on resource availability in terms of financial, organizational and technical resources. The *potential scope of OIS* can be defined in terms of a number of parameters which describe a particular office. These include: the size of the organization, the type of employees who work there, the level of support available, the design of the organization, its physical location and its context.

The organizational scan and the potential scope of OIS define two sides of the same coin. The organizational scan assists senior management in defining the constraints on OIS implementation; the definition of the potential scope of OIS estimates the potential requirements for OIS within the organization. These two efforts provide input to the feasibility studies described in Chapter 6. At this point they serve to assist management in comparing the requirements for OIS to what is feasible for the organization.

The Organizational Scan and the Feasibility Triangle

This assessment enables senior management to clarify the upper boundaries of feasibility for OIS in their organization. Naturally this involves both quantitative and qualitative assessments. By defining a feasible scope for OIS, senior management will be able to maintain stronger control of OIS development by defining better constraints. Note that this assessment is not a technical assessment, nor does it lead to system design. Rather it produces a set of organizational limits to the scope of OIS.

Development of an initial set of constraints is important since it focuses senior management's attention on what is feasible from an organizational perspective. This is particularly true for small organizations where an inappropriate application of OIS can be fatal (in North America roughly 70% of all employees work for organizations with less than 200 employees). In larger organizations the risk of financial collapse is less of a problem; however, the strategic use of resources (financial, technical and organizational) is still a major issue.

The feasibility triangle (Figure 2.1), illustrates how the feasible region of a possible OIS implementation is bounded by financial, organizational and technical resources. The bounded feasible region represents the limit of what is possible for the organization. All feasible OIS implementations should fall far inside this region.

It is important for management to establish the feasibility triangle before detailed proposals for OIS are developed. When this has been accomplished it will limit what can be a far-ranging debate to those alternatives which are inherently feasible. Furthermore, by addressing this issue first, one avoids the possibility of enthusiasm for technology overcoming good business practice.

Management can define a feasible region without any technical knowledge whatsoever (see the first half of the Minicase at the end of this chapter for

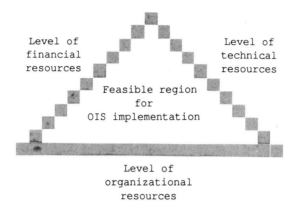

Level of financial resources

Level of technical resources

Feasible region for OIS implementation

Level of organizational resources

FIGURE 2.1   The feasibility triangle

an example). Certainly any competent financial officer can set the financial constraints based on the organization's current financial position and future commitments. The bottom line is the dollar value which, if exceeded, would result in extreme financial hardship for the organization. This is composed of two figures: first, the total cost of OIS which would shove the organization over the edge (in the context of other expenditures), second, the rate of cash flow which would take the organization to the edge. It is crucial that a senior manager have this boundary clear before OIS efforts begin.

Similarly, a competent general manager should be able to reach a shrewd assessment of the level of organizational resources available. This will involve an assessment of the ability of employees to adapt to the new technology, potential political problems and their likely impact, the general level of effort required by various individuals, and the physical environment necessary for effective performance. This 'soft' analysis is the key to ensuring that the organization does not attempt more that it can accomplish. For example, if heavy training requirements were associated with OIS, an organization which was already overworked might be unable to profit from the system.

Finally, management must assess the technical resources available. This may involve reports from in-house technical people as well as external assessments. In practice, we find managers too ready to delegate this technical assessment to in-house technical personnel. The current hardware and software situations are important, but the capabilities of the in-house information systems personnel are equally important. Technical skills and manpower are key issues along with the ability of IS people to communicate with the rest of the organization. This latter skill is important because an OIS ties the office together. For it to be successful there must be open, frank and mutually comprehensible communication between management, IS people and other employees.

In the following paragraphs the key issues associated with each category of resource are discussed further.

Financial Resources

In assessing financial resources, both capital cost and ongoing operating costs must be considered. In addition, the rate of cash flow and the entailed financial risks must be understood. The key issues here include the following:

(1) What is the total capital expenditure we can afford?
(2) What is the acceptable rate of capital expenditure?
(3) What degree of uncertainty is associated with these estimates?
(4) What is an acceptable level of financial risk for us?

In answering the first question, executives in smaller firms must be concerned with potential affordability problems. One simple way to answer the question is to determine the amount of capital expenditure that would put the company on the verge of bankruptcy. Presumably one would set the limits on OIS expenditures far below this level. In larger firms the issue of bankruptcy is of less importance. However, the strategic allocation of capital to various projects is always an issue. Here an executive must decide what an acceptable figure for capital expenditure is, given the alternative uses for that capital (one approach is value analysis, e.g. Strassmann (1985), ch. 8, pp. 136–150). Clearly this involves political considerations as well as financial ones.

As mentioned, capital expenditures have two components; the first component is the 'one time' capital expenditures necessary to start up or significantly improve the system. The second component is the 'ongoing' non-capital expenditures required for operating the system. Examples of each of these are listed in Table 2.1.

The second question is easier to answer than the first, since it is derived from current projections of cash flow over the next 2–3 years. It is important to forecast the likely rate of cash flow to determine the upper limits on the rate of expenditure for OIS in both short and longer terms. The last two questions address risk assessment. Each is important, and each can only be answered in terms of an individual organization.

Organizational Resources

These resources include the people, the organizational structure and the environment in which they operate. The key issues here are as follows:

(1) What is the extent and sources of possible resistance to change?
(2) What is an acceptable rate of change given our current environment? Are

TABLE 2.1 *Examples of capital and non-capital expenditures*

Capital:	Non-capital:
CPU	printer cables
keyboards	paper, forms
monitors	modem cables, phone cables
printers	power cords
PC cards such as:	PC cleaning equipment
parallel ports	monitor glare screens
printer cards	maintenance
modem cards	tapes
serial port cards	disks
network cards	non capital order expenses
memory cards	PC carrying cases
video cards	machine covers
software	printer ribbon, laser toner
main operating packages	software upgrades
standard office automation	user, professional training
material utilities ($>$ $250)	disk storage boxes, covers
mice	tape storage containers
joysticks	equipment/user manuals
hard disks, floppy drives	computer magazines
memory chips	immaterial equipment repair
delivery/installation charges	parts:
modems	tractor feed replacements
	knobs, switches,
	screws, bolts, bulbs,
	sheet feeders, paper
	trays, printer heads,
	fans, grills
	software utilities ($<$ $250)
Generally:	Generally:
anything that will contribute	anything that can
to the functioning of the	be easily destroyed
business for an extended time	or lost within a few
and is not likely to be easily	months.
destroyed or lost.	

    some departments more resistant to change than others? Why?

(3) What level of training is likely to be required for those involved in OIS?

(4) Given our current job pressures, will employees have enough time to participate in OIS design and in training?

(5) To what extent can we identify strategic areas for OIS?

(6) Do we have a climate that fosters change and experimentation? Do we coach for better performance or just punish failure?

(7) What physical changes do we have to make to our environment to enhance employee performance with the OIS system.

The seven questions presented here address the common organizational problems that we have encountered. Identifying likely resistance to change is important for OIS to be successful. While management cannot identify all potential sources, they can identify areas that may be most likely to resist change (we discuss this issue in more detail in Chapter 8). Furthermore, an assessment of an acceptable rate of change is important. If the organization has been through a major upheaval a cooling-off period may be necessary before engaging in an OIS project. On the other hand, introduction of OIS is used by many firms as an opportunity for organizational redesign. For example, organizations can be redesigned by reassigning job content and thereby redistributing areas of responsibility. The assessment of acceptable rate of change is clearly a subjective management decision.

Related to the problem of organizational change is the issue of training. If large amounts of training are required, or if employee time is very limited, the rate of OIS implementation may be severely constrained. Finally, organizations that foster a positive environment for learning and experimentation tend to have better success with OIS for the simple reason that people are less afraid to make a mistake. Use of OIS creates a potentially less constrained environment. Unless employees feel able to experiment and learn new ways to do their jobs, much of the benefit of OIS will be lost. (This idea is based on the concept of double-loop learning espoused by Argyris (1982) where the learning process focuses on the underlying assumptions and theories being used.) Consequently, management assessment of the climate is important. If changes must be made, make them before introducing OIS if possible. For example, Straussman (1985, ch. 9, p. 157) contends that well managed corporations tend to profit from computerization while poorly managed firms tend to suffer.

Technical Resources

These resources include those physical resources available to the organization (both internally and externally) and the technical expertise available to use those resources. The key issues to consider here include:

(1) Do we have sufficient technical expertise to conduct the OIS design and implementation in-house.
(2) Do our technical people have sufficient time to conduct the OIS design and implementation in-house?
(3) Does our physical location constrain access to qualified personnel and vendors?
(4) Can we build on our current technical systems or do we need to start from scratch with new hardware and software?

Review available MIS/OIS staff to see if they tend to formulate problems in business terms. If they respond with too much emphasis on technical solutions the needs of the business may not be well served. Another major danger is that management will be influenced by an overambitious MIS group who are unwilling to admit their own limitations. As we mentioned earlier, obtaining an outside assessment of technical staff and existing hardware and software may be a valuable option. At a minimum, management should review the track record of MIS/OIS personnel and assess whether or not they are up to the task of further projects.

The geographical location of the organization has a major effect on its ability to undertake OIS projects. If the business is located in a major city such as Toronto, New York, London or Paris, then clearly there will be few constraints in terms of vendors or availability of personnel. However, if the organization is located in a remote spot in a developed country (e.g. the Shetland Islands, Yukon Territories or Alaska) or in a developing country (e.g. Indonesia or Malaysia) then the availability of resources becomes a major issue. In particular, management must determine the level of vendor support available and the availability of qualified technical support personnel. If this is limited, then the number of feasible options for OIS development may be severely constrained.

Table 2.2 summarizes the key management questions associated with each of the dimensions discussed earlier. Once management have defined a feasible region for OIS implementation they can turn their attention to defining the potential scope for OIS secure in the knowledge that preliminary constraints on OIS implementation have been defined.

Defining the Potential Scope of OIS

The six office definition parameters listed in Table 2.3 help define the potential scope of OIS requirements. We show how each may be used to define the scope and feasibility for OIS in an organization (an example of the application of these parameters is found in the last half of Minicase 2.1 at the end of this chapter). During this discussion the pragmatic aspects of measurement are also discussed.

Parameter 1: Size

Size is perhaps the most straightforward of the office definition parameters. We have combined size with a taxonomy of office roles to give a more detailed picture of the office. The key issue here is to determine the potential demand for OIS hardware, software and training.

Size defines the overall requirements for number of workstations, communications ports, peripheral devices and space. Space requirements exist for the machinery itself, storage of accessory items (backup media, additional supplies), and work areas to perform off line repair and testing.

TABLE 2.2   Summary of key management questions for definition of feasibility triangle

Financial resources
1. What is the maximum capital expenditure we can afford?
2. What is the acceptable rate of capital expenditure?
3. What degree of uncertainty is associated with these estimates?
4. What is an acceptable level of financial risk for us?

Organizational resources
1. What is the extent and sources of possible resistance to change?
2. What is an acceptable rate of change given our current environment? Are some departments more resistant to change than others? Why?
3. What levels of training are likely to be required?
4. Given our current job pressures, will employees have time to participate in OIS design and training?
5. To what extent can we identify strategic areas for OIS?
6. Do we have a climate that fosters change and experimentation? Do we coach for better performance or just punish failure?
7. What physical changes do we have to make to our environment to enhance employee performance with the OIS system.

Technical resources
1. Do we have sufficient technical expertise to conduct the OIS design and implementation in-house.
2. Do our technical people have sufficient time to conduct the OIS design and implementation in-house?
3. Does our physical location constrain access to qualified personnel and vendors?
4. Can we build on our current technical systems or do we need to start from scratch with new hardware and software?

Given the breakdown by employee roles we can estimate the general demand for a variety of software systems and for training. Size has a direct impact on hardware and training costs, since each employee will require some level of support and training. For a small firm the management issues cluster around the problem of scheduling training so as not to disrupt work. Some small businesses are so lean that training time is scarce. For larger firms the issues centre around the size and composition of training groups and whether internal or external training programs should be used. In large multinational corporations, training may be highly formalized and allow little room for individual tailoring of programs (the issue of training is discussed in more detail in Chapter 8).

We suggest that a simple count of the approximate number of employees at each level will suffice for an overview of OIS demand. Assume that managers will generally want communication software and hardware, word processing, electronic mail, spreadsheet and easy-to-use database systems. Professionals and semi-professionals may want the same systems, but will likely need tailored

TABLE 2.3 Office definition parameters

Parameters which help define the type of office are listed here with suggested measures for each.

(1) Size
Number of:
  (a) Managers/executives
  (b) Professionals
  (c) Semi-professionals
  (d) Clerical
  (e) Secretarial
  (f) Support
  (g) Technical

(2) Type of people
  (a) Level of technical training
  (b) Orientation (internal/external)

(3) Level of support
  (a) Financial resources
  (b) Technical support (people and equipment)
  (c) People support (training and liaison)

(4) Design of organizational system
  (a) Decision making
  (b) Coupling
  (c) Augmenting and automating
  (d) Job design

(5) Location
  (a) Single office
  (b) Multiple floors, one building
  (c) Multiple buildings, campus
  (d) Multiple buildings, one city
  (e) Multiple cities
  (f) Multiple countries

(6) Context of office
  (a) Level of Risk
  (b) Degree of time pressure
  (c) Degree of concern for security
  (d) Organizational culture
  (e) Purpose and environment
  (f) Level of activity

software to support specialized analysis. Clerical workers will require carefully tailored systems to support their particular functions. Secretarial workers will require access to their boss's systems plus word processing, spreadsheets, and possibly time management software.

Support personnel, such as mailroom employees, copy department employees and others in similar jobs, will likely not be highly involved in using OIS. However, their activities should be given at least a cursory inspection to see if automation or augmentation can apply. Consequently, the number of people in this group may be an important datum for cost/benefit analysis.

The common denominator between these jobs is that as we go down the list from the executive worker to the support worker we move from *highly unstructured work* to *highly structured work*. This, in turn, implies that the degree to which we can define the activities of each employee and their interdependencies will increase as we move towards the structured work.

Consequently, the focus must be on systems which will be appropriate for the particular level of work to be supported. A system which permits numerous modes of inquiry, different modes of use, and which supports a wide range of functions may be suitable for unstructured work. A more structured and highly controlled system may be appropriate for more highly structured work. Whether this is a fundamental rule of design, or whether it is merely an artifact of our organizational thinking is as yet undetermined.

Once the number of employees in each category has been established, we have taken the first step toward defining the overall scope of the OIS. At this point we have a rough measure of the likely demand for a variety of software and hardware, and have preliminary data which will subsequently enable us to identify the likely level of training and support required.

Parameter 2: Type of People

The types of people available to the organization are differentiated on two dimensions; level of technical training or education; and orientation (internal/external). The key issue here is to determine the potential of different types of employees to use the capabilities of OIS.

Technical training   An individual's current level of technical training will have a major impact on the level of future training and support required. In general those who have some technical training or education will require less training and less support than those who do not. Furthermore, one can assume that they will be more likely to play with the system and learn new ways to do new things. Finally, technically literate people make ideal candidates for resource people who can provide informal support to their co-workers.

Orientation   The location of an individual in a network of internal and external communication links has a major impact on the type of systems these people will use. Employees with an external orientation are of particular interest. There are two types of external orientation.

Employees such as salesmen, agents etc. are *physically external* to the organization and require links to tie them more closely to the organization and to other employees. The use of portable computers, voice mail and electronic mail by these people provides the necessary organizational ties.

The second type of employee is *psychologically external*. In this instance heavy emphasis is placed on environmental scanning. These employees, who may be in marketing, purchasing or the board room, scan the external environment for information of use to the organization. They are usually physically located within the organization. From an information point of view, though, their main contacts are external to the organization. These people need access to databases, wide area networks, computerized bulletin boards, etc. to facilitate their environmental scanning. In particular this latter group of employees is often overlooked when planning an OIS.

Parameter 3: Level of Support

The level of technical support for hardware and software, and support for end-users, is critical to determining the level of OIS activity that is feasible for a particular organization. The key management issue here is to determine the potential availability and quality of support (both internal and external) for OIS hardware and software, i.e. there is the trade-off between hiring someone just out of university at $X per year, or someone with 5 years experience at $2X per year.

We noted previously that, in the more remote areas of many countries, there is limited technical support from vendors and limited availability of qualified people. In these cases technical support can become the limiting constraint in designing an OIS. In more central regions the issue of external technical support is less of an issue; however, internal support is always a concern. Costs for technical support taken together with training costs are often overlooked.

Required support for initial training, and ongoing liaison, are key elements in identifying the scope of the OIS project. Many consultants estimate that one should budget 50% of total hardware and software costs for training. Typically, organizations plan to spend 10% of total hardware and software costs on training. Not surprisingly, they find that their systems are not fully utilized. The direct costs of training are substantial; however, the indirect costs can be staggering as well.

In one small publishing firm located in New York, the only person who could operate the new telephone system properly became ill for two weeks. The owner estimated that by 'saving' $2000 on staff training they lost $25 000 in advertising revenue during the two-week period. One additional reason they avoided the training was that they had pared their staff to the bone. No-one had time for the training. The owner commented that we found time afterwards when we realized the costs of our decision.

Parameter 4: Design of the Organizational System

The design of the organizational system involves pondering the philosophy of the organization. This involves consideration of the organization's structure. The key management issue here is to determine how OIS will interact with the existing structure and culture.

There are many ways to organize a particular office. They vary from the highly structured to the highly unstructured. At one end of the continuum one finds the highly structured bureaucratic office; at the other end a variety of organic structures appear. In between, firms may be organized by function or by product line. While a discussion of organizational design is beyond the scope of this book, some parameters which we feel are key to the successful design of an OIS can be mentioned. For those who wish to pursue the concepts of organizational structure further, we recommend the book by Mintzberg (1979) in the list at the end of this chapter.

Decision making   The degree to which decision making is decentralized is a crucial variable, since a computerized system can both centralize and decentralize decision making. If the organizational fit is not correct, the result will likely be removal of the system and possibly the person(s) responsible for the system.

During the implementation of an E-mail system in a government agency, employees came to value the system because they could share experiences and problems with colleagues across the country. This led to a considerable amount of joint problem solving at the lower levels of the organization. Senior management, who had previously been central to this process, was not amused and eventually pulled the plug rather than take a peripheral role. (The study was conducted with a group of case workers at the Non-Medical Use of Drugs Directorate in Ottawa. An abbreviated account of the study may be found in Irving (1978).)

Coupling   Another key variable is the degree of coupling between various organizational units or workgroups. If organizational units are closely coupled (i.e. if a change in the inputs or outputs of one unit strongly affects the functioning of another unit), one must be very careful in designing an OIS.

An illustration of this concept at the macro level is the current move to computerized systems by police departments across North America. Police departments believe that information systems will make them more efficient at crime prevention and detection. However, the court systems have been slower to computerize. What then will be the effect of increasing the efficiency of a police force while not improving the court system? One police chief commented that, since the court system is overloaded now, increasing the efficiency of the police system could bring the courts to a standstill unless steps are taken to streamline procedures there as well. Clearly consideration of the wider implications of OIS are in order.

At the micro level the number of links a workgroup has to other workgroups indicates its degree of connectivity; the intensity of those links is a measure of the closeness of coupling to other groups. The key management issue here is the extent to which workgroups are interdependent and the effect of this interdependence on OIS design.

When data on workgroup links are combined with the current levels of technical support for workgroup tasks, management has information on the level of design activity required to develop an OIS. For groups with few links and little technical support, stand-alone systems may be appropriate. For groups with many close links and considerable support, a carefully planned needs analysis and system design procedure must be used. The links and their intensity also provide management with a checklist to identify individuals who must be consulted during the needs analysis and design phase of OIS. This issue is discussed more fully in Chapter 7.

Automating and augmenting   In adopting a strategy of automation the organization is often attempting to replace human effort with computer-based systems. In many instances it is creating a factory in the office. By adopting an augmentation strategy the organization is introducing systems which will assist humans, not replace them. In practice a mixture of the two strategies will likely be employed.

At the preliminary stages of OIS design it is a management responsibility to target areas for automation and augmentation and to facilitate discussion on the appropriate levels of each.

Job design   The issues of job design and redesign, ergonomics and level of employee participation in the OIS design process are always contentious. Current thinking is solidly behind providing an opportunity for substantive employee involvement in the design process (see Chapters 6 and 7 for more information). This becomes particularly critical if computerized performance monitoring (Chapter 9), flexible work schedules (Chapter 10), or end-user computing (Chapter 11) are to be introduced.

Parameter 5: Location

The physical location of offices can vary widely. *An office* may be in one location, or may be spread across several floors of a building, across several buildings in the same city, or across a wide variety of geographical locations. In many cases people who work together under the same supervisor are grouped proximally. However, in other cases, subordinates may be spread across North America or Europe. With the advent of telecommuting and remote work some 'offices' are conceptual or virtual. In other words they exist organizationally but not physically. The key management issue here is determining the degree to which the physical location of the office(s) determines the potential for OIS implementation.

The physical and temporal dispersion of employees has a significant effect on the demand for communication capabilities, the potential loads on these systems and the ultimate cost of providing these services.

Clearly there is less expense and effort involved in connecting employees located on one floor of one office building than there is in connecting the same number located in several cities. Even with employees on different floors in the same building there can be problems. Several years ago a firm located in New York found a crimp in their plans for OIS when they realized that the cable shafts in the multi-floor office tower they occupied were full. They could not get their cables from one floor to another. The problem was eventually solved, but at enormous expense.

Local area networks are appropriate for a single office, multiple offices in one building or for a campus layout. For multiple buildings in the same city one may consider dedicated data lines; for multiple cities a wide area network or a satellite link may be desirable. An example of the latter is the GEONET system employed by Manufacturers Hanover to connect its national and international operations.

International boundaries are a particularly complex problem given the globalization of markets. Moving information between countries must involve knowledge of their legal system. In general it is easy to move data between Canada, the USA and the UK. Germany and France restrict the options for international data transfer, and in some countries it may be illegal. Careful analysis of legalities of international information exchange is a must.

Parameter 6: Context of the Office

In terms of the context of the office we consider the level of risk that is acceptable, the amount of time pressure, degree of concern for security, the culture, the purpose of the office and the level of activity in the office. The key management issue here is the degree to which the context of the office may limit the scope for OIS implementation.

Risk   Pinpointing the acceptable level of risk is crucial. Financial risk is paramount, particularly for small organizations which must wrestle with the possibility of the system not delivering any benefits. For some organizations this would be disaster; for others an inconvenience. Personal and/or political risk is another aspect of risk that deserves consideration. If the organization is risk-averse, or if it is punishment-oriented, the OIS design will be more conservative than if the organization adopts a more entrepreneurial spirit.

Time pressure   The degree of time pressure is manifest in two ways: first is the amount of time available to design and implement the OIS; second is the amount of time pressure experienced by potential users of the OIS. In the first instance extreme time pressure means that a simplified, conservative design is necessary. In the second instance a sophisticated, simplified system that requires little learning is necessary.

Security   The degree of concern for security also defines the scope of the system design. At the primary level is concern for security of access. This is a technical matter; however, it is a management responsibility to ensure that it is addressed. Security from disaster or disaster avoidance is clearly a management concern.

Part of the system design must be the development of a disaster prevention plan. The key question to ask is: 'If our system were destroyed tomorrow what would be the consequences for our organization?'

Culture   In addition to other factors we must consider the collective beliefs and values (i.e. the organizational culture) of those who work in the office. Some organizations have a climate that is conducive to change and innovation. In these cases the implementation of OIS should proceed fairly smoothly. Other organizations are slow to change. Their culture is more conservative with respect to innovation. These organizations need change agents who are sensitive to the culture.

The example of electronic mail implementation mentioned earlier involved a group of social workers who were given an electronic mail system (this study was discussed under parameter 4). Not only did these implementors have to overcome the users' fear of technology, they had to deal with a culture that was decidedly anti-computer. They were able to overcome the resistance by spending time with individuals to help them learn the system, by encouraging individual innovation, and by demonstrating that the E-mail system facilitated individual jobs.

These same fears and concerns have been echoed by employees and managers since computers first began to invade offices. Argyris (1971) discussed managers' reactions to management information systems. Among the managers' responses were concerns about loss of control over their operations, worry about their flexibility in managing, and concerns that there would be an undue emphasis on narrow technical competence rather than more general managerial skills. Argyris noted that the technical people who typically implemented MIS (in those days) lacked the people skills to deal with these emotional reactions. Interestingly, the same observation holds true for many of these people today.

Identification of the characteristics of the people in your office helps define software requirements, training demands and level of support necessary. In addition it may help assess the likely political ramifications of installing the system. We discuss the operational details of how to do this in Chapter 6.

Purpose and environment   The purpose of the office is also a key consideration because it largely determines the context in which work is done. For example, offices such as claims-processing departments in insurance companies and large government offices such as Revenue Canada, or the IRS in the USA, are really white-collar factories. In these offices the main purpose is to process forms quickly and accurately. The work is highly structured, and tight productivity controls are maintained.

Other offices exist to facilitate professional work (consulting firms), monitor the work done in other offices (e.g. head offices), and conduct general management activities. Each of these offices has a distinct purpose and operates in a distinct environment.

Clearly the environment in a dynamic consulting firm is different from that in a stable, conservative financial institution. Government offices operate in a different environment from small privately owned businesses.

The consequences of these differences have major implications on the volume of communication and information flowing through the firm (information intensity), the time pressures on employees, and the levels of security and reliability demanded from support systems. For example, as the information and communication intensity increases, the size of the OIS required to carry the flow increases. As the time pressures increase, the design of the user interface increases in importance and the available training time decreases.

Level of activity   The level of activity in the office or the intensity of the work is also an important consideration, as is the seasonality of activity. The OIS system must be able to handle the peaks in activity. A typical example would be an accounting office. While the workload may be fairly constant much of the year, at year end there is peak of activity. Consequently, the seasonality of activity affects the timing of OIS implementation. The absolute level affects the availability of personnel for training and the rate of introduction of OIS.

Table 2.4 summarizes the key management questions associated with each of the six parameters discussed in the preceding paragraphs.

SUMMARY

In this chapter we have defined OIS and discussed some of the organizational parameters affecting the overall scope of the OIS effort. Obviously there are many issues arising out of this discussion. Some of the more important of these are outlined in Table 2.5. Each of these is discussed below.

Issue 1 focuses attention on the need to assess the organization and its functioning prior to considering OIS. It includes the possibility that organizational redesign alone may be sufficient to enhance organizational effectiveness.

TABLE 2.4   Summary of key management questions for determining the potential scope of OIS

Parameter 1: Size
What is the potential demand for OIS hardware, software and training?

Parameter 2: Type of people
What is the potential of different types of employees to use the capabilities of OIS?

Parameter 3: Level of support
What is the potential availability of support (both internal and external to the organization) for OIS hardware and software?

Parameter 4: Design of the organizational system
How will OIS fit with the organization's structure, culture and the decentralization of decision making? Issues such as automation or augmentation and job design should also be considered.

To what extent are individual workgroups interdependent and what is the effect of this interdependence on OIS design?

Parameter 5: Location
To what degree does the physical location of the office(s) determine the potential for OIS implementation?

Parameter 6: Context of the office
To what degree does the context of the office in terms of risk, time pressure, security, culture, purpose of the office and level of activity enhance or limit the scope for OIS implementation?

TABLE 2.5   Key management issues

---

(1) How can we improve the overall functioning of our organization?
(2) What is the appropriate balance between automation and augmentation given our industrial and organizational context?
(3) What is the feasible region for OIS implementations given our current levels of financial, technical and organizational resources?
(4) What is the potential scope for OIS implementation given our particular mix of people, technology, activities and environment?
(5) Given the feasible region and potential scope, should we proceed with OIS? If so, at what rate?

Assuming that OIS can contribute to improved organizational functioning, Issue 2 requires that management review their management philosophy and determine an appropriate balance between automation and augmentation. Clearly this review must be firmly grounded in the nature of their industry and the historical context of their office.

Issue 3 directs managerial attention toward the definition of a feasible region for OIS implementation. As we pointed out earlier, definition of the feasible region is *not* a technical issue, it is a managerial issue. Furthermore it is critical that management define a feasible region before engaging in detailed consideration of OIS.

Issue 4 focuses on the need to identify the potential scope for OIS implementation. This (need-driven) assessment is different from the feasibility assessment discussed previously. Here the focus is on potential application of OIS in the firm, ignoring the availability of resources. Furthermore, it is a non-technical assessment which requires management to 'ballpark' their OIS requirements in terms of hardware, software, training and support.

Issue 5 addresses the comparison of the feasible region to the potential scope for OIS implementation. If the potential scope is too far outside the feasible region it may be inadvisable to continue the process in the short term. If the potential scope lies far inside the feasible region perhaps a quicker implementation is possible.

Clearly these five issues raise a number of questions. The question of evolving an organizational structure which takes advantage of the flexibility of OIS technology is important. Another key question relates to inter-boundary efficiency. In other words, how can the organization avoid the problem of enhancing local effectiveness in one department while degrading it elsewhere.

Those questions, which are dealt with in subsequent chapters, are listed below in the order in which they are discussed in the text.

(1) To what degree should we integrate business processes? (Chapter 3).
(2) How can we reorganize to take advantage of the flexibility inherent in OIS? (Chapters 3 and 4).
(3) What is the strategic value of OIS to our firm? (Chapter 4).
(4) What are the consequences of OIS? (Chapter 5).
(5) How should we plan for OIS? (Chapter 6).
(6) How can we collect useful information on our requirements? (Chapter 7).
(7) How can we avoid inter-departmental inefficiencies? (Chapter 7).
(8) How should we approach the implementation and evaluation process? (Chapter 8).
(9) What issues should we be prepared to deal with in the near future? (Part III; Chapters 9, 10, and 11).

MINICASE 2.1: EXTRUSIONS UNLIMITED

Extrusions Unlimited Inc. (EUI) is a manufacturer of aluminum and plastic extrusions. They have two plants located in a major North American city, one for aluminum extrusions and one for plastics extrusions. During the past 10 years the company has grown from US$10 to $20 million in sales. Part of this growth was due to a new general manager, Bill Healy.

Though EUI has grown substantially, most of the improved productivity has been in the form of new production technology and methods. The production staff has increased by only 5%, from 200 to 210 full-time employees. During this period the office staff increased from five to 21 employees.

Bill is concerned that any future increase in business will overload the current office staff. His main option seems to be updating their current computer systems, which are to say the least outdated. Their accounting is currently done on a 10-year-old Tandy computer. They have just added 1 PC/AT in the marketing department to handle customer data.

As a first step Bill has completed an organizational scan or feasibility triangle. The results are as follows:

Financial Resource Limits

$150 000 total for hardware, software and training.
$20 000/year ongoing operating expenses.

Technical Resource Limits

All hardware support to be contracted out—turnkey system.
System must be tailored to EUI; no in-house programming.
Cut-over to new system must be done in July and August.

Organizational Resource Limits

No resistance likely—need recognized by staff.
Time for training is a problem.
Training must be extensive—low skill levels now.
No more than two staff in training at a time, due to workload.
Cross-training is crucial since staff cover for each other.
Training must not be more than one day per person per month.

Bill set his $150 000 limit based on the company's current cash flow, his projections of likely future cash flows, and estimates of current and planned capital projects. This limit was constrained since the company was extended already due to the updating of the plastics extrusion line.

EUI had little in-house computer expertise. Though the engineers were computer-literate, Bill reasoned that their time was too valuable to be used on OIS. Consequently he felt that the design, support and training should be contracted out.

Since the current staff had been with the company for some time and working relations were good, Bill felt that there were few if any political problems to worry about. However, given the heavy workloads in the office, he was concerned that there would be enough time for training.

Once Bill had his feasibility triangle defined he felt comfortable bringing others into the planning and design process. The next step was

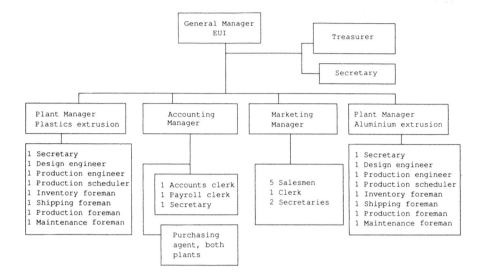

FIGURE 2.2   *Organizational structure for Extrusions Unlimited. Notes: (1) Plastics plant is 10 miles from aluminum extrusion plan; (2) all administrative functions are at Aluminum plant (General Manager, Treasurer, Accounts Manager, Marketing Manager)*

to independently identify the potential for OIS based on the general characteristics of his organization.

The problem Bill faced was to determine the potential scope for OIS based on his assessment of organizational needs. Bill and the marketing manager reviewed the organization. Their preliminary analysis was as follows:

Parameter 1: Size

Managers	6
Professionals	5
Semi-professionals and supervisors	15
Clerical	3
Secretarial	6

Bill estimated that three of the six managers would need some level of computing power, and that all of the five professionals would need support. The secretaries clearly needed word processing. The semi-professionals (which included the salesmen and the foremen) had little need for support now. The clerical staff likely required some support, probably in the form of spreadsheets. In all 17 workstations of some type would be required. Bill estimated that this would cost in the neighbourhood of $50 000–70 000 for the workstations alone.

Parameter 2: Type of People

Bill and his marketing manager reviewed the capabilities of their staff. They felt that the four engineers would likely require minimal training. On the other hand, the remaining 13 employees had little computer expertise and clearly required extensive training and support.

Bill was concerned about the salesmen and their need for support in the field. In addition the marketing manager expressed an interest in linking to external industry databases.

While they were unable to estimate training accurately they guessed that it might cost at least $20 000 including time off the job.

Parameter 3: Level of Support

They had estimated hardware and training costs at a total of $70 000–90 000. Clearly other costs, such as software and support costs, would be involved, as would costs for furniture, air-conditioning, etc. They decided to assume that these costs would be equal to the hardware costs as a first approximation. The final approximation put the demand for financial resources in the range of $120 000–140 000.

They then turned to the issue of technical support. At present EUI had no official technical support personnel. One of the engineers helped out if there were problems. Once they went to a more ambitious OIS clearly both hardware and software support would be required. They reserved judgement on whether or not this support should be provided by an internal employee or by contracts with external suppliers. For internal support they estimated that a technical support person would cost a minimum of $50 000/year. Furthermore, since the training demands would be heavy the first year they felt that a trainer would be required for that period. Since EUI was in a central, urban environment they were sure that there would be no problem in locating qualified personnel or in obtaining technical support from a variety of potential vendors.

Parameter 4: Design of the Organizational System

Bill felt that the size of their organization precluded extensive automation, with the exception of payroll and accounting functions. Clearly the engineers required specialized software, as did the marketing groups. Furthermore, Bill wanted a more open structure than they had previously. One way to achieve this would be to tie everyone together on an electronic mail system and/or a voice-mail system.

One troublesome problem with the current organization was that the current structure and administrative mechanisms had evolved on an *ad-hoc* basis. Bill was determined to review their current structure and improve it during the process of implementing OIS. He was particularly concerned that they avoid developing computer support for ineffective systems.

Parameter 5: Location

Location was generally not a serious problem for EUI since most of the administrative work was done at the aluminum plant. One concern was the type of connection that would be required to the plastics plant. Bill was unsure of the technical details, however; he felt it was important that all eight employees at the plastics plant be tied electronically to head office at the aluminum plant. His reasoning was that even one error or omission caused by lack of communication could be very costly in a service-oriented business such as his.

Parameter 6: Context of Office

In terms of organizational context the main concern Bill had was financial risk. EUI was debt-heavy due to plant expansion, and funds were tight for the next three to four years. Bill estimated that costs for OIS would have to be carefully monitored. Time pressure was another concern since EUI was relatively lean. Most employees were heavily booked and had little time to learn a new system. Security, except for financial information, was not a major concern. The organizational culture was supportive, largely as a result of Bill's effort over the past few years to build a positive, team-oriented environment.

Parameter 7: Workgroups and Links

The workgroups at EUI were tightly linked with each other within each plant; links between plants were weaker. Marketing was tightly linked with the production group at each plant. Bill wanted to strengthen these links as well as improve coordination between the plants. He had heard recently that shipping from the plastics plant was having difficulty coordinating with shipping from the aluminum plant on joint orders (those containing both plastics and aluminum extrusions). He wondered if OIS could strengthen the flow of information here.

Balancing the Organizational Scan and OIS Scope

Bill soon realized that the proposed scope for OIS was very close to his upper limits for cost. His main concern was whether or not they could scale back the OIS and still obtain major benefits. Clearly some careful planning was required. Bill decided to proceed to a planning and needs analysis phase, but decided to keep tight management control of the process.

Before proceeding to the next phases, Bill formulated his concerns as follows:

(1) Which key activities in my firm most need support?
(2) What key communication links must be maintained?
(3) How can I keep this project focused on business issues?

(4) Who should be involved throughout the design process?
(5) Should I consider a major organizational redesign?

How would you answer these questions in the context of your firm?

BIBLIOGRAPHY

Ackoff, R. (1967) 'Management misinformation systems', *Management Science*, **14**(4), B147–B156.

Argyris, C. (1971) 'Management information systems: challenge to rationality and emotionality', *Management Science*, **17**(6), B275–B292.

Argyris, C. (1982) 'The executive mind and double-loop learning', *Organizational Dynamics*, Autumn, pp. 5–22.

Baetz, M. L. (1985) *The Human Imperative: Planning for People in the Electronic Office*, Holt, Rinehart & Winston of Canada, Toronto.

Hirschheim, R. A. (1985) *Office Automation A Social and Organizational Perspective*, Wiley, Chichester.

Huber, G. P. (1984) 'The nature and design of post-industrial organizations', *Management Science*, **30**(8), 928–951.

Irving, R. H. (1978) 'Computer assisted communications in a Directorate of the Canadian Federal Government: a pilot study', in M. C. J. Elton, W. A. Lucas and D. W. Conrath (eds), *Evaluating New Telecommunications Services*, Plenum, New York, pp. 455–469.

Mintzberg, H. (1979) *The Structuring of Organizations*, Prentice-Hall, Englewood Cliffs, NJ.

Strassmann, P. A. (1985) *Information Payoff*, Free Press, New York.

Tapscott D., Henderson D. and Greenberg M. (1985) *Planning for Integrated Office Systems: A Strategic Approach*, Holt, Rinehart & Winston of Canada, Toronto.

Tapscott, D. (1982) *Office Automation*, Plenum, New York.

Uhlig R. P., Farber, D. J. and Bair, J. H. (1979) *The Office of the Future*, North Holland, New York.

Chapter 3
OFFICE TECHNOLOGY AND THE INTEGRATION OF BUSINESS SYSTEMS

To place the rest of the book in its proper context it is important to understand the underlying concepts and terminology indigenous to the technology and industry surrounding office information systems. Until recently a manager's office would have included a desk and chair, the size and elegance of which would reflect the owner's status, a single- or multi-line telephone, a desk calculator, and an assortment of legal pads, pens, pencils and other stationery. A secretary seated outside provided clerical services. The manager's calendar was a small notebook which was carried in a pocket or briefcase so that the individual was always 'in touch' with his or her schedule. The office of today's manager is often radically different. The imposing desk has been supplemented or replaced by a personal workstation, the calculator is now the size of a credit card—and maintains a phone directory as well; electronic mail has replaced the internal memoranda, the calendar is maintained on a portable personal computer, and a voice-mail system is in place to answer the phone. In addition, the various corporate information systems may well be interconnected so that information located in a different city or country can be accessed with the same ease as data located in the user's hard disk. This chapter explores office technology and the issues related to interconnecting this technology.

INTRODUCTION TO OFFICE TECHNOLOGY

The magnitude of change that has occurred in office technology in the past decade is remarkable. Sifting through the confusion is a difficult task for the uninitiated. This section explores some of the major technologies available to today's managers, and explains the key skills required to master this myriad of systems.

Computer Systems

Modern computers vary dramatically in size. One of the most common categorizations of systems is the distinction between microcomputers, minicomputers and mainframes. While recent hardware advances have greatly blurred the performance distinctions between these systems (some micros today have as much power as minicomputers, and minis are fast challenging mainframe performance) the computing industry has retained the categorization as a way of roughly differentiating their systems. In fact the driving factor in choice of a system is the complexity and type of application or programs the system will be required to run. This brings into consideration the speed with which the system can process transactions and access data, versus the price the manager is willing to pay. Keep in mind that many systems currently in use today are a result of historical system investments, even though the same ability is now available in a cheaper and more efficient form. Other criteria becoming increasingly relevant in decisions on hardware purchases include the degree of centralization, security and control management wishes to have over the information to be captured, compatibility, and communication ability. Linking less powerful machines versus buying one large machine to service all needs is becoming more and more feasible and cost-effective. Linking also serves to provide a backup for one system if the other fails; however, at some point a degradation in performance will justify the purchase of a more powerful machine in the next 'class.'

Microcomputers are the smallest machines. They are designed for use by single users or the 'knowledge worker' (as opposed to minis and mainframes which support more than one user at a time). A recent significant change, however, is the ability to run multiple applications on a single PC, allowing the user to easily transfer and use information between different programs. Hardware characteristics include a central processing unit or CPU (the brain of the machine), disk drives, a monitor and a keyboard, all of which sit on top of a desk. The main memory of microcomputers (often referred to as random access memory or RAM), which is the part of the CPU that holds programs when they are running, can range in size from 640 000 bytes (640K) to 8MB (where MB represents a megabyte and is approximately 1 000 000 bytes). Secondary memory, used for storing information even when the computer is turned off, consists of the floppy and/or hard disk drives. Floppy disks come in several varieties; $5\frac{1}{4}$-inch floppies were one of the earliest form to be developed. Double-sided double-density $5\frac{1}{4}$-inch floppies hold 360K bytes of information. High-density $5\frac{1}{4}$-inch floppies hold 1.2MB. More recently, $3\frac{1}{2}$-inch floppies have been gaining popularity. The diskette surface is better protected by the hard casing than the older flexible casings. Regular density $3\frac{1}{2}$-inch diskettes hold up to 712K bytes, while the high-density diskettes hold 1.44MB. Hard disks are much larger, typically holding from 20MB to 40MB, although 300MB and larger sizes are also available.

Minicomputers are slightly larger (although physically the smallest will fit on a desk top or under a desk). They are designed with a departmental focus with respect to an organization. They usually support from 20 to over 200 simultaneous users and provide both greater storage capacity and greater main memory and processing power. Main memory typically ranges from 32MB or 64MB up to 512MB. Secondary memory, in the form of a hard disk, holds from 280MB to 622MB. Typically there is more than one hard disk, or 'disk pack', attached to minicomputers to provide larger amounts of storage. Note that the capacity of a minicomputer can overlap with microcomputer capacity on the low end and mainframe capacity on the high end.

Mainframes are designed for an enterprise focus or for applications where data are shared and used all through the organization. They are typically installed in their own specially designed rooms. An air-conditioning system is required to keep the computer from overheating. In addition to the processor, one or more disk packs (like disk drives only much bigger) and tape drives are also stored in the computer room. Users access the system from remote locations using some form of workstation (see below). Users do not maintain their own physical disks for a mainframe system; rather, they access a portion of the disk pack which has been set aside for them by the systems staff. Mainframes can support a large number of simultaneous users and offer still greater storage and processing power than the minicomputer class. Indeed, while the hardware RAM and disk pack sizes can be similar to the minicomputer (although they can also come in larger sizes), a mainframe allows for greater capacity to add these features together. In practice, one will find widely differing opinions as to where the line is drawn between a mini and mainframe computer.

Workstations

Several types of workstations are available to managers. *Dumb terminals* allow the operator access to a centralized mainframe or minicomputer system, but do not have any processing capability of their own. These systems consist of a monitor and a keyboard only. No floppy diskettes are used with these systems. *Stand-alone personal computers* consisting of a microprocessor, a monitor, a keyboard, one or more floppy diskettes and perhaps a built-in 'hard drive', are another commonly available workstation. These stand-alone PCs do not need to be connected to any other workstations. The software is stored on the hard drive or the floppy diskettes.

Networked personal computers represent an alternative to standalone PCs. The individual workstations are the same as stand-alone PCs, but much of the software and data that is commonly used throughout the office is maintained on a separate 'server' machine, and is shared by all of the users. Linking the PCs via a network also allows the users to communicate electronically and to share files. *Intelligent workstations* are another variation of workstations. These consist of a personal computer (as described previously) which may or may not be part of a network, but which is linked to a central computing system and can operate as a terminal for that system. These workstations are called 'intelligent' because they include a processor other than the processor of the central system.

The monitors available for each of the above configurations also vary. Many monitors display a single colour only. These *monochrome monitors* may be in black and white, black and amber or black and green. *Colour monitors* typically support many colours (256 colours are available on some) and many shades of those colours. Monitors also differ in the *resolution* of images on the screen. Resolution refers to the number of dots of light per inch. By increasing the number of dots per inch a more precise representation of the image can be achieved. *Graphics support* is another feature available with some computer monitors. Graphics capability allows the user to display pictures and graphs as well as standard text.

Printers

A variety of printers are available for office systems. These printers vary in the quality of output and the speed of operation. Two general classes of printers are available—impact and non-impact printers.

Impact printers create characters on a page by striking the surface with some device containing the image to be transferred, in the same way that typewriters create characters. These printers differ in terms of the device which creates the characters. *Dot matrix printers* form characters by combining dots to form the required patterns. The number of dots used to form each character, and thus the quality of the output, depends on the number of rods or pins which make up the print head. Inexpensive dot matrix printers use nine pins to form the characters, while more expensive printers use 12 or even 24 pins. Dot matrix printers produce output at speeds of around 100–200 characters per second (CPS). Near letter-quality printing can be achieved on a dot matrix printer by printing over each character more than once, with the print head moved slightly to the left or right of the first pass. Thus, the spaces between the dots are filled in by subsequent passes. *Daisywheel printers*, which are becoming rare, print one character at a time, using preformed characters on a print element which strikes the page to produce the character. These letter-quality printers produce about 55 CPS.

Non-impact printers do not rely on the print head striking the printing surface in order to create characters. *Ink jet printers* spray ink onto the paper in the form of the desired character. They typically operate at speeds between 90 and 200 CPS. *Laser printers* differ from all of the preceding printer types since they print an entire page at a time, rather than forming each individual character separately. An inexpensive laser printer will produce about six to eight pages a minute, while a high-output laser printer will produce up to 600 pages a minute. Laser printers are a great deal quieter than impact printers.

Recent developments in copier technology have created a new class of printers called intelligent copiers. *Intelligent copiers* provide a combination of functions which characterize both printers and photocopiers. One advantage is that they save floor space while providing multifunctionality.

Applications

The most common applications available for managers are word processing, spreadsheets and databases. Each of these will be discussed in turn.

Word processing systems are powerful tools to assist in the preparation of written documents. From simple memos to reports to entire books, document processing can be greatly aided by the use of word processing systems. Revisions and corrections to documents do not require retyping of the entire document, so they can be made much more easily and quickly.

Early word processing systems required the user to insert complicated codes for formatting into the document. Thus, the screen viewed by the user bore no resemblance to the finished product. Most currently used word processing systems show the user exactly what will be produced on paper when the document is printed. In these WYSIWYG ('what you see is what you get') systems, the user can see, at any given moment, what the finished product will look like. Formatting is controlled by menus which can be called up as needed. Controlling formats by menus frees the user from having to remember all of the complex codes to activate various features.

In addition to facilitating the task of editing a document, word processing systems support several advanced features which aid in document processing. First, most systems include a dictionary and a spell-checking system which can be activated to check an entire document or give the spelling of a particular word. Some systems even provide a phonetic speller which suggests possible words based on phonetic similarity rather than character similarity. In addition to spelling assistance, some word processors provide a thesaurus. For long reports with many sections a word processor can generate a table of contents from the headings in the text. Indexes on certain key words can also be created. Figures and tables can be inserted into documents with automatic referencing. The primary advantage of all of these advanced features is that additions or changes to the document do not force the user to go back and renumber tables, or recreate indexes and tables of contents. Rather, all of these changes are handled by the software itself, freeing the user to concentrate on other matters.

Desktop publishing is a new area of word processing which is becoming more commonly available. Desktop publishing software is essentially a very powerful system which allows the user to combine text, graphics, and charts on a single page, using multiple columns and a wide variety of fonts and styles. Furthermore, by adding optical scanners to the system, photographs and pictures can be reproduced on the computer and incorporated into the document.

Spreadsheets, such as Lotus 1–2–3, are another category of commonly used office applications. Spreadsheet packages are typically used in financial analyses, such as budgeting or decision support. For example, consider the development of a budget, both manually and using a spreadsheet package.

To create a budget manually the user might begin by projecting sales for three years. If sales are expected to increase by 10% each year over the current year's sales, then current sales are multiplied by 110% to calculate next year's sales, and so on for three years. Expenses might also be projected to increase by a fixed amount, and the necessary calculations would be made for each of the three years. Alternatively, expenses might be projected as a percentage of the year's sales. To calculate expenses, then, the projected sales is multiplied by the percentage expected, and each expense account is thus derived. Once all of the sales and expense figures are calculated, the expense figures are totalled, and subtracted from expected sales, yielding the expected profit.

With a spreadsheet package the process of entering a budget is somewhat different. Often formulas are entered instead of numbers, and the system calculates the result. Thus, in our budget example above, the sales figures would be entered as 'CURRENT SALES FIGURE * 110%' and the system calculates and prints the result. The same is true for each of the expense categories, and for the profit calculation as well. Allowing the spreadsheet to do the calculation achieves two ends: first, mathematical errors, or errors in transposing the results, are eliminated; second, conducting sensitivity, or 'what if' analysis, is greatly simplified. Any of the figures in the spreadsheet can be changed, and the rest of the spreadsheet will be recalculated to reflect that change. Thus, if the sales projection is raised to reflect new market information, all of the expenses that are represented as a percentage of sales will be recalculated on the new sales figure, and the profit calculation will be redone—all without intervention by the user. Compare this to the manual case, where each of the expense calculations must be redone and then the totals changed to show the difference.

Once the completed spreadsheet is entered, several additional features can be invoked. Graphs can be created to show particular features of the projection. For example, expenses can be represented in a pie chart, to demonstrate where the bulk of expenses are incurred. Similarly, sales can be graphed over time in a line chart, showing the expected increase (or decrease) in sales. In fact, any information that is represented in the spreadsheet can be displayed in graphical format. Statistical analysis can also be conducted using spreadsheet software. In the budget example, for instance, statistical comparison of budgeted versus actual figures might be conducted on a monthly basis to determine the accuracy of the projections. Over time, such analyses would contribute to the ability to predict sales and expenses, thus allowing the organization to better manage its financial situation.

Like word processing systems, most spreadsheet packages are menu-driven; thus, the user does not have to remember the complicated steps involved in creating a graph or printing a spreadsheet. Rather, the menu system moves through the functions step by step prompting the user for options as it goes.

Database applications are another powerful tool which can assist the user in managing various aspects of the organization's information. Many larger databases are maintained by a special functional group within the organization. This group is responsible for identifying all of the relevant data to be maintained, and storing it in a format that is both efficient and flexible. The efficiency aspect of database storage refers particularly to the ability to store information without redundancy.

For example, consider a firm which distributes office supplies. Prior to the evolution of database systems, information about customers would be maintained in several places. The shipping department would maintain information about orders received from each customer, and the address to which the product is to be shipped. The accounting department would also maintain information about customers, such as billing address, invoices billed and payments received. If, as was the case prior to database systems, this information was maintained separately by the individual departments, a change in a customer's address would have to be made by both departments (and any others that maintained customer information). Quite often updates would not be made to all files containing the information, or would be made at different times. Thus, invoices could be sent to the wrong address, causing delays in payment. In a database system, however, information about customers is maintained in one central location, and is accessible to the various departments as they require it. Thus, when the customer's address is changed, a single change to the database will ensure that all of the departments have the correct information.

Flexibility is the second important goal of database systems. In the pre-database era, information was maintained in the format that was required by the particular application for which it was gathered. In the preceding example, for instance, product information would have been maintained according to the customers who had ordered it. This organization of data is fine, as long as information about total sales of product X are not required. In order to calculate that information, each order record would be scanned to see if product X was part of the order. If so, the number of units would be added to the total, and the next record checked, and so on. More complicated queries—like the number of units of product X sold to customers in a particular geographic region—would be even more difficult to process. In a database system, by contrast, the data are stored in such a way as to facilitate the processing of *ad-hoc* queries such as these.

While database systems are often managed by a central function, individual users can gain access to the database to update information and process queries (within the limits of the security imposed by the system). In addition, portions of the database can often be downloaded to a personal computer for additional processing by an individual.

Local Area Networks

Local area networks (LANs) provide a link between stand-alone personal computers and the applications which run on them. LANs allow several users to share software, data and peripherals (such as printers), and to communicate electronically with one another. Several varieties of LANs (topologies) exist.

Star networks use a single, dedicated, computer as a network control unit. Each component on the network (including workstations as well as peripherals) is directly linked to the control unit and sends messages to that unit when it requires access to another component—to print a document or send mail, for example. The control unit then provides the necessary service for the requesting unit.

Bus networks consist of a single cable (the bus) with several cables linked into it to support the various network components. Messages are sent directly to the device to be accessed when required, rather than routing the request through a controller as in the star network. The difficulty with bus networks is that each component has to compete with the others for access to the bus. If too many messages are being sent across the bus at once, errors can occur or performance speed can drop substantially.

Ring networks are a final variation on network topologies. In a ring network each of the components is connected to two others, forming a closed loop (or ring). When access to a device on the network is required, a message is sent, through one of the neighboring components. When a component receives a message which is not directed at itself, it passes the message along to its next neighbor, and so on until the message reaches its destination. In practice there is a tendency in many offices to designate one of the PCs on the ring as an 'official' server.

LANs provide substantial benefits to the users, as noted above, but these benefits are not without cost. All things being equal, software performance (response time) is higher when the software is not shared. Furthermore, some software cannot be installed on a network, since it was designed for single users only. The LAN itself also requires considerable maintenance and day-to-day management. Someone has to be responsible for adding workstations, diagnosing and solving simple malfunctions, backing up (archiving) important programs or data and otherwise ensuring that the performance of the network is acceptable. Thus, it is important to weigh the costs and benefits of a network carefully before proceeding.

Optical Storage

One area of change in computing and communications technology is a move towards optical storage technology. *Optical disks* record information by having a laser burn tiny spots into a disk. This method results in much more compact data storage than is available with magnetic media such as floppy disks. Currently optical disk technology in *CD-ROM* format (compact disk–read only) can store 550MB of data. These systems are similar to those used for audio CD drives. Currently the price of these systems is decreasing (a drive is around $800) and the market is growing. CD-ROM is useful for material that does not require frequent revision (e.g. encyclopedias, dictionaries). As the price declines they will be used more for storage of permanent records. The second type of drive is the *write once, read-many drive* (WORMS). These drives allow the user to write on the compact disk one time only. Once written the information cannot be revised. However, since these systems usually store a gigabyte (1024MB) the average user can use one disk for a long time. The drives are currently very expensive, but are expected to play an increasingly important role over the next five to ten years. The final type of optical drive is the *erasable optical drive*. This is leading-edge technology and none are currently available for the mass market. If one were available at an affordable price it would revolutionize information storage.

The major implication of optical storage is that, in the very near future, vast amounts of information and data will be available to the average PC user. Students will access a VideoDisc encyclopedia instead of a book. In a few years students might receive optical disks when they begin school at grade one, and use it to record all their work for the rest of their lives. Much of what is currently printed on paper will be stored on non-erasable optical disks. As a manager, one use of this technology is clearly going to be for archival purposes.

Advanced Telephone Technology

Only a few years ago the thought of a detailed discussion of telephone technology in a book on computer systems would have seemed a bit absurd. Similarly, the need to attend telephone training sessions would have seemed totally outrageous. However, the advances in telephone technology made possible by the advent of digital switching and other telecommunications advances have vastly extended the functionality and complexity of the office telephone.

The human voice is carried via waves of sound. Early telephone technology was designed to carry these sound waves across some distance to a receiver on the other end of a connection. An analog (wave) signal was used to carry these sound waves. This configuration is ideal for voice transmission; however, it leaves a great deal to be desired if the goal is to transmit data.

Computers store data as electrical impulses. A series of 1s and 0s (representing the presence or absence of a signal) represents all of the information stored by computers. These digital signals must be converted to analog signals if they are to be sent across a traditional telephone line, using a modem (which stands for *mo*dulator–*dem*odulator). This extra step in sending digital signals slows the transmission of data. In addition, noise or static which has little effect on our understanding of sound transmissions can cause errors in the transmission of data, since the noise could easily alter a single digital impulse.

The move to replace the analog technology with more convenient digital equipment has increased the accuracy and speed of data transmission. *ISDN*, or integrated services digital network, is a developing standard or 'master plan' for creating a fully digital telecommunications network which can transfer voice, data, facsimile, image, and video communications through a common communications channel. This global system would also link public and private networks and provide conversion between incompatible hardware systems, thus connecting users worldwide to any other system. The promise of ISDN is great; however, we can expect it to be many years before its full potential is achieved.

While the full potential of digital communications is far from a reality, several advances have found their way into the marketplace.

Cellular Telephone Technology represents a real breakthrough in telecommunications. By use of overlapping frequency cells, users can maintain a relatively clear signal while on the move. This allows a businessperson to be in constant voice, fax and computer communication while in transit; thus making use of 'dead time.' New developments, such as satellite-based phone systems hold the promise that, by the turn of the century, an individual can be accessible anywhere on the globe.

Voice-mail systems use specially designated hardware and software to store a digitized version of a voice message. The user can retrieve messages by dialling a particular number which activates the retrieval system to play back the message. In addition, messages can be forwarded to other addresses, to circulate important information to other users.

Facsimile transmission (FAX) systems use optical scanners to translate text or images into digital or analog signals which are then sent across a telephone line to a similar machine on the other end which re-creates the image from the incoming signal. The ability to transfer documents electronically is a tremendous advantage for fast-moving businesses.

PUTTING THE PIECES TOGETHER

The preceding sections have described individually the components that make up typical office systems. In this section the implementation of these technologies for three hypothetical offices is illustrated: the first is a small private law office; the second is a medium-sized accounting firm, with three partners and several assistants; the third is the sales office of a national consumer goods manufacturer. Each of these offices requires a different configuration of office technologies to meet its particular business needs.

The Law Office

Bill Brown opened his private law practice in 1988 following a merger of his former employer with another large law office. He handled mostly corporate clients—small businesses with fairly simple legal requirements. His practice was located in a fairly large city, in the main floor of a home, the remainder of which had been converted to apartments. Bill employed an assistant who provided secretarial and administrative services.

The computing equipment that Bill purchased for his practice included two stand-alone personal computers—one for himself and one for his assistant; a dot matrix printer for himself, and a laser printer attached to his assistant's computer. In addition, he had purchased a cassette recorder and a transcriber for taking dictation, a small office copier, and a telephone answering machine.

Word processing was the primary application on the computers. All of the correspondence with clients was maintained on the computer system, as well as legal briefs and other documents. Bill did a fair bit of his own writing on the computer, but usually turned the documents over to his assistant for final editing. Often he would dictate letters, have his assistant enter them and then edit the final printed copy. This allowed him to prepare work in the evening, have it typed during the day and review it the next evening.

Bill also used a spreadsheet package for the accounting records of the business. Each month he would record expenses in a spreadsheet. At the end of the year the expense totals for each category could be transferred to an income statement spreadsheet. Billable hours and chargeable expenses for each client were also maintained in spreadsheets to facilitate the invoicing process. Once Bill had completed a job and recorded all of the expenses, he would print a copy of the spreadsheet to a file and give the diskette to his assistant. She would import the printed spreadsheet into an invoice shell in the word processing software. Then she attached the standard covering letter and printed the package for mailing.

Overall, Bill was satisfied with this relatively simple system. He had considered the possibility of networking his and his assistant's computers, but had decided that the additional cost of managing the network would more than offset the cost of duplicating the software. Furthermore, since the assistant did most of the final printing, only her machine needed access to the laser printer. Bill is currently considering purchasing a low-end FAX machine to speed up billing and correspondence with clients.

The Accounting Firm

The accounting firm of Taylor, Beckett and Chen was founded in 1979 by Allison Taylor. Fred Beckett joined the firm in 1981 and Eleanor Chen was persuaded to join in 1986. In addition to the partners, the firm employed two associates, four accounting students and three secretaries.

All three of the principals were heavy computer users, but Eleanor had experience with programming as well. On her advice the firm hired an associate in the summer of 1987, who would also act as part-time systems manager. The partners had computers prior to that, but these were badly in need of upgrading. The systems manager undertook a comprehensive study of the activities of the business and finally recommended a system of networked personal computers. Four portable computers were also purchased for visits to client companies. A dedicated file server maintained most of the firm's applications (word processing, general accounting both for the firm itself and its clients, spreadsheet, and a small database of client files). Each of the partners' machines also had hard disks on which they maintained personal applications. For instance, Eleanor liked to maintain her personal calendar electronically so she had a small personal productivity management system on her machine. Fred often gave presentations and liked computer graphics, so he maintained a presentation graphics package on his machine. Allison had recently installed a desktop publishing package on her computer. She felt that professional publishing would improve the presentation quality of her consulting reports. The remainder of the machines did not have hard disks.

Several printers were located throughout the office. Each of the partners had a personal near-letter-quality (NLQ) dot matrix printer. In addition, an ink jet printer and a laser printer were attached to the network for higher-quality printing. The firm also has a high-speed FAX machine which is used heavily for communication with clients. In addition most partners have a FAX at home.

The systems in place in the accounting firm were somewhat more complex than those in the law office. A part-time systems manager was required to manage the network and otherwise coordinate the systems details. The use of networked PCs allowed better document transfer, software sharing and communication, while the hard disks on the partners' machines allow them to maintain applications of interest only to them, thus reducing the workload on the network.

The Sales Department

The ACME Widget Company was a national manufacturer of a line of consumer goods. Their regional sales offices handled $60 million annual sales to distributors and large chains. The central regional sales office, managed by Terry Dutton, employed a staff of 30 managers, salespeople, administrative assistants and secretaries. In addition to a new telephone system with voice-mail capability they have a high-speed FAX and a more sophisticated computer system than do the other two firms.

The sales office was linked to the central computing system of ACME through a leased line (a phone line leased full time from the phone company, i.e. a dedicated line). This allowed the office to be fully connected to the central computer (a mid-sized mainframe) as if the mainframe were on the premises. The mainframe was used for order entry, customer database enquiry, budget and control systems, and communications with other managers (both at head office and in the other branches).

Most of the staff used dumb terminals to access the central computing systems. Their needs were limited to corporate systems, and thus the added expense of personal computers was not required. Terry and the secretaries, on the other hand, had personal computers which could also be used to link to the central computer.

The secretaries had never liked the mainframe word processing package. They felt that the microcomputer software for word processing was much better, and that they could produce better documents faster using a micro-based package. Terry did very little of his own word processing, but was a frequent user of presentation graphics software; he too preferred to use a PC. The three PCs were networked to provide access to the laser printer. A daisy wheel printer and a high-speed band printer in the sales office were linked to the mainframe system.

Systems management was mostly provided by the corporate head office, since the regional offices had little on-site hardware. A consultant in the information systems department was assigned to the central sales office to provide support when the department required it. A help line in the information centre was also available if the staff had questions about particular software packages.

Summary of Introduction to Technology

The preceding mini-cases presented the major office technologies that a manager could expect to encounter. The advancements in computers and communications technology virtually ensure that this discourse will be out of date within a short period of time. However, the understanding of basic technologies which you should now have will help you comprehend the changes that will certainly occur. In the next section we discuss the trend toward increasing integration of systems that will characterize the next decade of computing.

INTEGRATION OF BUSINESS SYSTEMS

The *integration* of computing and communications systems is a powerful force in the design of new technologies and systems. The development of the *ISDN* standard is one factor in this integration. It will allow transmission of voice, data, image, facsimile and video communications across the same communications channel and will be faster than current technology. Theoretically, when ISDN is implemented worldwide, computers in any country will be able to talk to computers in any other country. Of course, the human language barrier will limit the ability to communicate across national boundaries, but the technology to do so will be in place nonetheless.

Other forms of software and hardware integration are also appearing in the computing industry. Computer manufacturers are attempting to provide better integration between mainframe, mini and microcomputer systems. For example, the same software interface would be used by applications on all three machines. In environments where more than one of the machines was operating, the user would not have to know which of the machines stored the required data. The systems themselves would handle this step in the processing.

Most writers on OIS discuss the importance of systems integration at the technical level. However, when one examines the implications of an 'integrated office system' it is obvious that the issue is a complex one that goes beyond specification of technical standards. Indeed, the key issue is the integration of business systems—how far—how fast, and why. The general management issue is:

TO WHAT LEVEL SHOULD ORGANIZATIONAL PROCESSES AND TASKS BE INTEGRATED?

Three Levels of Integration

Integration can occur at three levels: *physical integration, operational integration* and *logical integration*. An additional aspect is *transparency*, or the degree to which the integrated system is invisible to the user.

Physical Integration

Physical Integration refers to the physical combination of previous separate elements. Northern Telecom's DisplayPhone was an example of physical integration, since a terminal and a telephone were combined in one physical unit. Combination telephone, alarm clocks, and am/fm radios represent physical integration on a more prosaic level. However, just because two things are combined in one plastic case does not mean that they can operate together. For example, while the telephone and the clock are in the same unit, their circuits are separate. There is no way the operation of one will substantially affect the operation of the other, barring a malfunction.

There are many examples of physical integration in OIS. Perhaps the simplest is the RS232 connection. This 25-pin connector is an industry standard.

Virtually all printers have an RS232 plug; however, just because you can plug your printer into your personal computer, minicomputer or mainframe does not mean that you can print text. The 25 pins can be coded in a variety of configurations. Consequently, one requires a configuration table or translation table for the printer and for the computer to which it is connected. Once the translation tables have been obtained, and the printer set-up performed, operational integration has likely been achieved. At one time this would be accomplished manually by adjusting the dip switches on the printer; now it is done through software-controlled programs. To the extent that one must still locate the specifications for a particular printer and 'set-up' that printer for the hardware/software configuration used, the system is still only at a level of physical integration.

Operational Integration

Operational integration refers to the capability for joint operation of two distinct entities. In the example of a tele/clock/radio, the telephone is only physically integrated with the clock and the radio. However, the clock may be operationally integrated with the radio. For example, on most clock/radio sets you can set the clock to turn on the radio at a certain time. This is operational integration since the operation of the clock affects the operation of the radio. The clock sends a signal to the radio and the radio 'understands' this signal as an 'on' or 'off' command. Presumably, the radio could be connected to turn the clock on or off.

Returning to our example of the printer, once the translation has been established, the printer will be able to perform most tasks involving the printing of standard text and alphanumeric output. However, this does not mean that one can print any output produced by the computer. Certain software programs may have unique output or require special graphics packages that your printer lacks. One example occurred recently with the use of a laser printer with Wordperfect. The particular laser printer has software for most of the Wordperfect output but lacked the upper and lower left corners of the bold line, in the line draw mode. Not a major lack to be sure; still it caused irritation and generated paperwork, and time loss for the person in charge of such matters.

Logical Integration

Logical integration takes the process one step further, in the sense that the operation of one entity is comprehensible to the other. In our example the clock and the radio are not 'logically integrated' since the radio does not 'understand' what time it is and the clock does not 'understand' what program the radio is playing or know what station it is tuned to. However, imagine that voice recognition technology has advanced to the stage where it is both cheap and simple to recognize continuous speech. Through the wonders of modern technology (vintage 2001) you now have a functionally integrated tele/clock/radio beside your bed. You can program not only the time but a range of interests into the beast. For example, since the radio understands the clock to the extent that it knows what time it is, and since the clock can interpret the continuous speech of the announcer, you can program the clock to switch on the radio and check the weather. If it's raining then the clock will turn off the radio for an hour; if it's sunny then the clock will tell the radio to turn up the volume until you get up and shut it off—all for $99.99, available anywhere!

From the perspective of an OIS system, logical integration implies that:

(1) every hardware device connects to and understands signals from every other hardware device;
(2) every software system provides output in a form that is usable by every other software system;
(3) all software functions properly with all hardware.

In a logically integrated system a printer would correctly print output from any computer and any software located on that system. We are still some way from 'off-the-shelf' logical integration. However, most mainstream software packages have some capability for producing and accepting output acceptable to other systems. For example, Wordperfect will accept files from other word processing packages; Lotus produces output acceptable to Wordperfect, etc. Despite these achievements, the user still has much to learn before the system becomes easy or natural. This ease of use, or transparency, is the ideal toward which OIS systems must strive if they are to provide major benefits.

Transparency

Transparency is the degree to which the integrated system is invisible to the user. Perhaps the most common example of system transparency occurs in learning to drive a car. When you first begin, you have to deal with a confusing welter of information. However, as you become more familiar with the process and the technology of driving it becomes easier. Finally, it becomes so habitual that you are scarcely aware of the manual actions required to drive and can concentrate on the traffic around you. At this point the technology of driving is transparent to you—it does not interfere with your concentration on getting from point A to point B.

Transparency explains why it is difficult for computer experts to explain a system to a novice. The former have likely reached a level where the whole system is transparent to them while the latter find that everything is both new and complex.

In a completely transparent OIS the user would not know where data were located and would not have to; the system would know the location. In addition, text output from a word processing package could be combined with graphics from a spreadsheet program by simply inserting the graphics where desired in the document. To use a database program would require simple English commands, and so on. A fully transparent OIS would require no technical knowledge on the part of the user. Assuming that one understood the business and the tasks to be accomplished, one could proceed without expending time and energy to 'learn the system.' Clearly, we have some way to go before achieving full system transparency. However, it is possible to use technology to support the integration of various business processes and tasks once we know what we want to integrate and to what degree.

Identification of 'What' to Integrate

In addressing this issue, managers must decide how to identify where integration will occur and what level of integration is desirable in each area. Note that this is primarily a *management* issue, since we are concerned with integrating business processes. In other words, we are examining the organization and the closeness of coupling between various elements and deciding whether or not to change the *status quo*. Once these business decisions have been made, integration becomes a technical issue.

From a business perspective there are two distinct categories or areas to consider: tasks and roles. Integration of tasks refers to the interconnection of various processes. Examples include the integration of a transaction processing system for order processing with an automated inventory system. Integration of roles refers to the combining of hitherto distinct organizational functions or jobs. This could be combining the jobs of warehouse clerk and order entry clerk, or the combining of production engineering roles with design engineering roles.

Integration of Tasks

In any organization there are a wide variety of tasks and processes. Some of these, such as order processing, may currently be done either by people (manually) or by computer systems. Others, such as text processing, may be accomplished by a combination of computers, other machines (typewriters) and people. Depending on the level of computerization an organization has at a given point in time, and how far they are prepared to go, the effort required to integrate a set of tasks can vary tremendously.

Tasks can be classified according to their degree of structure. Unstructured tasks include most work involving interpersonal communication. Many successful managers operate in an unstructured way; communicating briefly throughout the day with a wide variety of people on numerous topics.

Kotter (1982) notes the gap between concepts of management behavior and actual behavior. *Management behavior* is conceived to be planning, controlling staffing, organizing and directing. *Actual behavior* is long hours, fragmented episodes,and oral communication. Kotter (1982, p. 156) goes on to state that:

> Actual behaviour, as a study of successful general managers show, looks less systematic, more informal, less reflective, more reactive, less well organized and more frivolous than a student of strategic planning systems, MIS, or organizational design would ever expect.

While the tasks described by Kotter cannot be integrated, the support technology can. We hear a lot about DSS (decision support systems). A more general concept would be USS (unstructured support systems)—systems to support unstructured work.

USS generally include communications systems which permit individuals to communicate with a wide range of people in a maximally flexible way. While managers and professionals seem to be obvious candidates for USS, management must consider the possibility of their use at lower levels as well. Unstructured support systems include teleconferencing (audio and video) systems, computer mail and conferencing systems and voice-mail systems. In addition, there are the wide area networks and the local area networks which provide support for a wide variety of potential communications.

Structured support systems support work or processes which are (or can be) highly structured. Whether or not some processes should be highly structured is an entirely different issue. In some cases these structured systems may take over the work entirely. For example ATMs (automated teller machines) have completely replaced much of the work of a human teller.

Photocopiers, FAX, wordprocessing and similar systems seem to fit as either structured or unstructured support systems depending on how they are used. It is less important to correctly classify them than it is to recognize that these systems can be used in a variety of ways.

It is important that organizations differentiate between relatively isolated tasks and mainstream processes that cut across all areas of the organization (in Chapter 7 we show how to do this using the HIT methodology). It is here that many failures occur, and it is here where organizations can potentially reap the largest gains. We discuss this issue in the next section. The management issue that must be addressed is:

WHAT DEGREE OF COUPLING SHOULD EXIST BETWEEN TASKS?

Integration of Roles

One issue management must consider is the possibility of redesigning the organizational structure. This can involve redesign at all levels of the organization from individual jobs, through workgroups, to the level of departments. For example, introduction of a computerized system for handling insurance claims may tremendously reduce the role of clerical help in the organization. As more tasks are put on line the roles of data entry clerk and file clerk may disappear, to be replaced by customer service representatives who input and retrieve data from a terminal.

Introduction of online word processing and communications capabilities in an academic department of a university may change the nature of secretaries' jobs. As more and more professors compose written work online, their jobs may change from entry-level typing to being an editor and specialist in the formatting of text. In this case automation of tasks (text processing) results in a change in organizational roles as well.

The major issue for management is to decide at what level to address the design problem. Based on our experiences over the past few years it seems that the workgroup is the natural starting point. A workgroup consists of a supervisor and his or her immediate subordinates. Furthermore it is the place where critical organizational tasks and roles can most readily be studied. However, when doing this level of analysis one must take care to identify crucial task links with other organizational groups. Once the tasks, the inter-workgroup links and the existing technology are identified for one or more workgroups the following managerial issues can be addressed:

WHICH ROLES IN EACH WORKGROUP ARE MOST SUBJECT TO TECHNOLOGICAL SUPPORT?

WHEN REDESIGNING THE ROLE AND TASK STRUCTURES IN EACH WORKGROUP, WHICH OTHER WORKGROUPS MUST BE CONSULTED?

When these questions are answered across all workgroups the foundation for the design of an integrated system has been laid. In particular, one will avoid the common problem of making one workgroup more productive at the expense of another. In Chapter 7 we show how to collect the data required to address these issues.

A Management Perspective on Technical Issues

While it is tempting to relegate technical issues related to systems integration to technical experts, there are a number of related management issues that must be addressed. Though we do not expect general managers to be experts on the technicalities of systems integration, we do expect them to be aware of the issues and ensure that the technical experts are addressing them. These issues include:

(1) standardization of hardware;
(2) standardization of software;
(3) standardization of communications;
(4) standards for systems integration;
(5) documentation of the whole process as listed above.

For an office system to be completely integrated, the first four issues must be addressed to the level where anyone in the organization can interchange text, data and images with anyone else. Furthermore, each employee must be able to process or manipulate the text, data and images (graphics) received from any other person in the organization.

This latter issue is a non-trivial one. For example, many organizations use a number of 'what you see is what you get' word processing systems (WYSIWYG—pronounced wissywig). Unfortunately, these systems all use different conventions for page breaks, headings, etc. And these commands are imbedded in the formatted text. Consequently, a document formatted by Wordperfect may be unintelligible to a user who has a different WYSIWYG. Naturally, there are special software packages which can convert one output to another. However, these packages are not completely reliable. There are often conversion problems even going from one version of a WP package to another version.

From a manager's perspective it may not matter which particular hardware or software system is adopted as long as key organizational tasks are supported. However, the manager should ensure that his or her technical support staff have developed plans to standardize hardware, software and communications protocols. Furthermore management must ensure that a plan is in place which will integrate hardware, software and communications into a seamless whole.

Though most executives will not understand the details of the technical issues, they must clearly understand related business issues. Consequently, managers should have at least the same level of knowledge of computing and telecommunications terms as they do of finance and accounting terms. In other words, they may not be experts in the field; however, they must understand key concepts and how these apply to their business.

Unfortunately, we are not yet at a stage where universal standards have been adopted. The integrated services digital network standard is still a decade away from being fully implemented. Furthermore, even though most vendors have adopted the X.25 standard for connectability for public networks, there are many dialects in existence. Just as an American, a Canadian and a British person may have difficulty understanding each other, so these networks can be made to talk to each other, but it is not easy. As a result, management should be careful not to underestimate the difficulties inherent in any systems integration initiative. Integration for the sake of integration benefits no-one.

This leads to the necessity to provide a business focus to all OIS activities and to IS management. If technical managers are not oriented properly to the business concerns, they will be early adopters of new technology because they have a fascination with it. Functional management, on the other hand, will have a primary fascination with running their areas of business effectively. These two domains of interest may be at odds. Consequently a functional manager may find it necessary to curtail the adoption of new technology if it will not integrate easily with current systems.

The main management issues pertaining to technical systems integration are:

WHAT MANAGEMENT POLICIES MUST BE DEVELOPED TO ENSURE APPROPRIATE LEVELS OF SYSTEMS INTEGRATION?

WHAT UNDERLYING COMMUNICATIONS ARCHITECTURE IS APPROPRIATE TO SUPPORT OUR FUTURE SYSTEMS DEVELOPMENT REQUIREMENTS?

The contrast between GM and Toyota illustrates the distinction between a technology-oriented system and a business-oriented system. When GM realized that it had to move toward a 'Just-in-time' inventory system to reduce inventories, it spent millions developing an integrated computer system so that production forecasts could be shared electronically with suppliers. The suppliers were forced to match their systems with that supplied by GM. When Toyota opened its new plant near Toronto it installed a system that accomplished the same goal by using FAX machines to communicate with suppliers. They realized that the key *business* goal was to communicate production information with suppliers.

SUMMARY

This chapter has provided a very brief introduction to the OIS technology and concepts of system integration. A summary of the management issues is listed in Table 3.1. We have discussed key hardware and software technologies and distinguished between different levels of system integration. In addition, issues relevant to management have been raised. The concept that management must assimilate is that integration of business systems is a long-term and arduous process involving considerable expense and not a little risk. In order to create a viable system you must start with a viable systems architecture and derive the hardware and software specifications from it. The basis for the communications architecture must be firmly rooted in the nature of the business itself, and consequently must be driven by business needs rather than technical expediency. In other words, the business strategy must drive the systems architecture, not the other way around. In order to facilitate this

TABLE 3.1   Summary of management issues

(1)   What are the relevant types, configurations and abilities of technology and how do they relate to general business concerns?

(2)   To what level should organizational processes be integrated?

(3)   What degree of coupling should exist between tasks?

(4)   Which roles in each workgroup are most subject to technological support?

(5)   When redesigning the role and task structures in each workgroup, which other workgroups must be consulted?

(6)   What management policies must be developed to ensure appropriate levels of systems integration?

(7)   What underlying communications architecture is appropriate to support our future systems development requirements?

process, one must understand the strategic implications of OIS technology and the technical implications of business strategies. This is discussed in Chapter 4.

BIBLIOGRAPHY

Cheong, V. E. and Hirschheim, R. A. (1983) *Local Area Networks*, Wiley, Toronto.

Foley, J. D. and Van Dam, A. (1983) *Fundamentals of Interactive Computer Graphics*, Addison-Wesley, Don Mills, Ontario.

Housley, T. (1979) *Data Communications and Teleprocessing Systems*, Prentice-Hall, Englewood Cliffs, NJ.

Keen, P. G. W. (1988) *Competing in Time*, Ballinger, Cambridge, MA.

Kotter, J. P. (1982) 'What effective general managers really do?', *Harvard Business Review*, November–December, pp. 156–167.

Senn, J. A. (1990) *Information Systems In Management*, 4th edn., Wadsworth, Belmont, CA.

Tannenbaum, A. S. (1981) *Computer Networks*, Prentice-Hall, Englewood Cliffs, NJ.

Chapter 4
STRATEGIC USE OF OIS

INTRODUCTION

There has been substantial discussion among MIS professionals and academics regarding the strategic use of computer technology. The discussion has focused at various times on computers in general (Porter, 1985a, b), computers in the plant (Warner, 1987), and on information systems as strategic weapons (Grindlay, 1983). Other authors such as Keen (1988) have focused on the strategic development of a communications infrastructure or architecture. We focus on the strategic implications of office information systems.

In order to place our remarks in perspective, the chapter begins with a discussion of the concept of using technology for competitive advantage. We then turn our attention to strategic use of information systems technology. Once the background has been developed we focus on the issues surrounding the strategic use of OIS.

USE OF TECHNOLOGY FOR COMPETITIVE ADVANTAGE

Though the idea of using office technology to obtain a competitive advantage is fairly new, production technology has been used for centuries in this manner. Consequently, many of the concepts have been applied in other venues. For example machinery has always been viewed as a means to increase productivity and reduce labor costs. In this section we discuss these general concepts of the strategic use of technology. We then discuss the unique strategic issues generated by OIS. Determinants of industry attractiveness and competitiveness are examined first, followed by a discussion of generic strategies.

Determinants of Industry Attractiveness and Competitiveness

There are a number of factors which contribute to the attractiveness of an industry and to the level of competitiveness that one may find there (Porter, 1985a). Among these factors are:

power of suppliers;
power of buyers;
threat of new entrants;
threat of substitutes;
rivalry among existing competitors.

The *power of suppliers* to determine how attractive and competitive an industry can be is a function of the degree of control that suppliers have over the particular industry. For example, Bell Canada and AT&T each had a monopoly on the supply of telecommunications services. The power of these suppliers was very strong, and the level of competition was naturally very weak. However, once deregulation was accomplished in the USA, and partial deregulation was a fact of life in Canada, the power of the two monopolies was reduced. Consequently, the industry itself became attractive to a number of potential companies and the competitiveness of the industry increased. Today, one can receive bids from a variety of companies on PBXs and in-house telephone systems.

There are a number of other industries, of course, where suppliers of a service can affect how attractive an industry is and how competitive it is. If everyone has to buy from one supplier, obviously the degree to which people compete and the way in which they compete, will be different.

The relative *power of buyers* can also affect industry attractiveness and competitiveness. For example, in an industry with a wide range in the size and power of buyers a few large buyers can control the price of goods for the smaller buyers by controlling or pressuring the suppliers.

For example, in the produce industry, a few large supermarket chains have substantial influence on the wholesale price of vegetables and fruit. They can get these products at a much lower price than can the small operator because they control huge volumes. Thus there is an unequal level of competition. The small buyers cannot obtain goods as cheaply as can larger firms. They must compete by reducing their overhead or by increasing their level of service (e.g. staying open longer hours).

Another factor which determines industry attractiveness and competitiveness is the *threat of new entrants*. An industry may be very attractive, but if it is easy for anyone to enter then it will become very competitive as new entrants are attracted to it. Eventually, increasing numbers of new entrants will drive down profit margins and the industry will stabilize.

In response to this threat a number of industries have established artificial entry barriers. For example, cab companies are licensed. In New York a medallion costs on the order of $70 000 and there is a lineup of people to get them. This is an artificial entry barrier since anybody with a car could potentially operate a taxi cab.

In some cases the barriers to entry are inherent to the industry. The steel industry is a good example. It is extremely expensive to set up a new steel mill and consequently there are only a few mills and they tend to be large operations.

In office automation the entry barriers are low and consequently there have been many new entrants. There are a huge number of small computer systems available, and a large number of software companies providing a wide variety of products. In addition the office systems market includes telephone consultants, telephone product suppliers and office furniture suppliers. Consequently there is a tremendous diversity of potential suppliers of equipment to the office. In these environments an office manager may find that it is hard to know what to buy and from whom to buy it.

Another factor which determines industry attractiveness is the *threat of substitution*. For example, in the computer industry many suppliers are making their personal computers IBM-compatible so in fact there is a high degree of substitutability between products. IBM is hurting because people are buying IBM clones and look-alikes at about half the price of an IBM PC. IBM's recent move to their PS2 and a microchannel architecture is an attempt to reduce the substitutability of their products.

A final factor is the *level of rivalry* between existing competitors. In the computer industry there are traditional rivalries between IBM, DEC, Xerox and a host of smaller competitors. In part these rivalries may not be based on logical or hard-nosed financial considerations. Some of them may be the result of long-standing competition, similar to that which might exist between two hockey teams or two baseball teams.

Given that there are a number of determinants affecting industry competitiveness, it is imperative that senior management assess strategic success factors for their industry and their organization. In other words, they must know how they are influenced by their suppliers, how they are influenced by their customers, the potential threat of new entrants, and what the potential is for substitution. In addition they must assess existing industry rivalries.

Three Generic Strategies

Once a firm has completed an assessment of the level of competitiveness in its industry it can begin to consider the strategic options open to it. There are three fundamental or generic strategies that any organization can use. One common strategy is to be a *low-cost producer*. In other words, to produce the product at the lowest possible cost, possibly sacrificing something in quality or service to the customer. Eastern Airlines, is an example of this strategy.

A second generic strategy is to produce *highly differentiated products*. These are products which are distinguished (at least in the mind of the purchaser) from those produced by other competitors. For example, consider the difference between Volvo and BMW. Volvo traditionally differentiate their products on quality, safety and reliability. Though BMWs are certainly reliable and safe, they tend to differentiate more on the status–performance dimensions than do the Volvos.

Finally, one can adapt a *focused strategy* which tries to identify some particular niche in the industry that is not covered now. The strategy is to fill that niche before somebody else does. Typically a firm identifies a vacant niche and develops a product. The firm then charges top price until other entrants are attracted to that niche. At that point, it can adopt a strategy of becoming a low-cost producer by continually lowering its price as competitors come in hoping to drive it out. This is exactly what Northern Telecom tried to do with their displayphone. They were one of the first in the market with a combination telephone and video display terminal. They started out with an extremely expensive product and, as other people came into the market, they continually lowered the price.

STRATEGIC USE OF INFORMATION TECHNOLOGY

In order to highlight the generic strategies discussed in the previous section, we discuss each strategy in terms of information systems technology. Subsequently we present and discuss current thinking on the strategic use of OIS technology.

Information Systems and Generic Strategies

Low-cost Producer Strategy and Information Systems

Information technology as part of a low-cost producer strategy has a number of effects: first, it can reduce the number of clerical staff by performing the tasks which these people do; second, it can permit better utilization of facilities and resources. Careful use of information technology can also allow significant reductions in inventories and accounts receivable by improving an organization's ability to analyze and control these areas. Finally, it can (if properly applied) provide better utilization of materials and lower wastage. The benefits to management in terms of having better decision making and better control over expenditures, and being able to correct errors quickly, are obvious.

It is vital to recognize the consequences of adopting this strategy. If a firm is to be a successful low-cost producer its goal has to be reduction of the unit cost per transaction to as low a level as possible. In the case of computer technology, this means that the firm must adopt a *bytes for bodies* strategy, which replaces people with computer software wherever possible.

One industry where being a low-cost producer seems to be a requirement is the insurance industry. We have interviewed several insurance companies: without exception, senior management stated that the only way to make money in insurance industries is through service, and by cutting down the time required to pay claims. In order to do that, most insurance companies have adopted an aggressive strategy of computerization to reduce their clerical staff to the minimum number. To some extent this has worked. However, when insurance companies have reduced their staff to the minimum level and reduced their costs about as far as they can go, and given the best service possible, the basis of competition will shift.

Product Differentiation and Information Systems

The strategy of differentiating your products from those of your competitors is a tried and true one. It has been used in a wide variety of industries. Information technology can help you differentiate your product in a number of ways. First, information technology can be a significant component of the product itself, or it can provide a service that is somehow unique and consequently can help attract consumers to your product. Information technology can also help differentiate products by significantly reducing the lead time for product development, customization or delivery. It can facilitate the modification of a standard product to meet a customer's special needs. It may also give a higher level of customer satisfaction. The following examples highlight these efforts.

In the insurance industry there is continual development of new kinds of consumer products. We have recently seen nonsmokers insurance and nondrinkers insurance, and a variety of dental insurance plans, a variety of medical coverage, travel insurance and apartment insurance. All these new products represent an attempt by insurance companies to develop a product which is a little different from their competitors. Some products may be a consequence of a focused strategy, which we discuss later; others are just variations on a theme. One company may, because of some unique software system or in-house organization, be able to offer a significantly higher level of customer service. Or it may be able to tailor the outputs to some particular form that its customers want. As the business gets more competitive, this capability of differentiating products on the basis of service, or on the basis of customization to customer requirements, is going to become even more important.

For example, a number of banks and investment houses are now offering a variety of investment services and portfolio management systems that run on personal computers. This allows customers to contact their investment agent and get a minute-to-minute update on how their portfolio of investments is doing, make changes, and get better information more quickly. A number of banks both in Canada and the US have introduced a service where you can find out how your investments are doing or make investments over the phone.

An early example of the use of information technology to differentiate a product occurred when an investment house in New York became the first to offer its clients the ability to trade in London, the US and as far east as Hong Kong, by having a series of computer links with the markets in those countries. In effect, sitting in New York, I could contact my broker at 5:00 a.m., do some trading in the London exchange and I could keep on trading until possibly 11 p.m. I could buy gold on the London exchange, and sell it in Hong Kong the same day. However, once one producer was able to differentiate its product, all the other producers had to develop similar products in a very short period of time, just to stay competitive. Now 24-hour investment services are commonplace.

A Focused Strategy and Information Systems

Using information technology in support of a focused strategy is different from its use in the other two generic strategies. To be able to focus on a particular niche in a market one needs good information on the total market. Consequently, using information technology to tie into external information databases can help the organization identify special market niches.

An organization can also use information technology to produce information intensive products or services. One simple example of this is the use of computerized PBXs to generate detailed accounts of telephone use by number called, by the extension number used, by the duration of the call, by use of a variety of long distance calling lines, and by use of special features. This kind of system can produce the statistical information automatically through a computerized recording on a monthly basis and is used by many companies to control and manage their telephone use.

Management Issues and Generic IS Strategies

Given the preceding discussion, it seems clear that managers must understand their competitive environment, which strategy is best for them, and how information technology fits with that strategy. If the effective way to compete is to adopt a differentiated product strategy, and the firm adopts a low cost producer strategy, it is headed for trouble.

Consider the management of personal trust funds. Most fund managers do not take anybody unless they have at least $100 000. Consequently, if you adopt a low-cost producer strategy you may in fact be eliminating the type of customer service these clients demand. In fact you might be better off increasing your overhead somewhat to be able to provide a higher level of service. On the other hand, trying to differentiate your product on a basis of increased service when the only way to compete is to be a low-cost producer is also unacceptable.

Consequently, the management issue related to the generic information systems strategies is:

WHICH MIX OF GENERIC STRATEGIES BEST FITS OUR INDUSTRY, OUR WAY OF DOING BUSINESS, AND OUR CLIENT BASE?

Anyone can identify a strategy and adopt it; doing so successfully is another matter. Consequently, it is important to understand, in so far as possible, how to use IS technology successfully, and what traps exist for the unwary.

Successful Use of Information Systems for Strategic Advantage

Curley (1984) indicates that an important first step in the successful strategic use of IS is to tie the introduction of technology to achievable and desirable corporate goals. Organizations must be willing to experiment with office technology, and must link that experience with the setting of objectives. By using the technology in a variety of ways they should be able to use their corporate learning to set more realistic objectives. Curley suggested that pilot projects and post-audits are more meaningful than cost–benefit analysis. It is important to remember that the costs and effort involved in these projects must be commensurate with the usefulness of the results expected. However, trying to identify all the possible costs and benefits may prove fruitless when many of both the costs and the benefits may not be directly expressible in monetary terms. It is better to develop pilot projects, monitor them carefully, and conduct thorough post-audits to find out what happened and why it happened. In this way the organization can enhance its learning potential and its creative ability to use the technology wisely.

Managers must actively support the learning process necessary to take advantage of the technology. They need to create an environment where people are judged on their batting average, not on whether or not they made a mistake. This strikes to the heart of many of the management problems because many managers do not assume a coaching or supportive role, but take a judgemental role toward their employees which discourages experimentation.

Frohman (1982) suggested that one of the key factors in the successful competitive use of technology was having a majority of managers in the business with some technical education and work experience in the organization. One needs both. Managers must understand the company and must have sufficient technical education so that they understand what is possible with the technology and are comfortable with the technical issues that may arise. Where technology has been used competitively, the managers tended to have a good technical background. Managers should also allocate funds among projects that will support and maintain technological leadership in specified areas. In most cases it is not possible to be a technological leader in everything. The corporation and the managers who work in it must be able to target those areas where it is worthwhile maintaining technological leadership. They must ensure that funds and support are available.

The decision making systems and structure of the company also tend to reinforce the priority given to technological matters. First, the firm's decision making systems provide a close connection between a business and technological decision making. As the level of technology increases, more information is available for the company to make decisions. This creates a closer link between technology and decision making. Second, the systems for decision making on technological matters should be consistent with the company's other decision making procedures. Huber (1984) presents a model for the post-industrial society where management of decision processes will become the dominant organizational paradigm. While Huber may overstate his case, there is no doubt that the increased information flows generated by a fully integrated office will demand more attention to decision processes than is common today.

Porter (1985a) provides an industry perspective on the competitive use of technology. He suggests that organizations analyze the information intensity of their products and of their industry. On one hand there are industries which are information-intense, such as the finance or the insurance industries. On the other hand, some other industries such as steel making and lumber may not be information-intensive. In addition, one should look at the role information technology plays in the particular industry. Is it mainly a support role, is it central to the functioning of the industry (a production role), or is it a peripheral role? We can identify and rank the ways in which information technology may create competitive advantage if we understand the role of information technology and how information-intensive the industry is. Another approach is to investigate how information technology might help develop new businesses and to develop a plan for taking advantage of information technology.

Grindlay (1983) summarized some of these approaches. He identified six ways in which IS technology can be used as a competitive weapon. It can produce a *defensible entry barrier*. The existence of a well-developed, sophisticated information technology may enable a company to defend itself against entry by possible competitors. Technology can be used to *strengthen customer relationships* by raising the 'cost' of switching to new suppliers. You may, for example, give a pharmacy a terminal to connect it to your drug company (e.g. American Hospital Supply).

Information technology can *change the intra-industry competitive balance*. The fact that some companies have successfully used information technology may put them in a much better competitive position (e.g. American Airlines SABRE System). A group of companies that use the technology wisely may find themselves basically competing with each other and being able to shut out smaller competitors. Information technology can *change the basis of competition from price to value-added*. Information technology also allows an organization to better *manage its supplier relationship* by monitoring quality, delivery dates, prices, etc. Finally, the technology itself can be used as a product or service to *develop new markets* (e.g. Merril Lynch's Cash Management Account). During the process of developing new products or services for an existing market, the companies may create new markets for their products.

Dangers in Using IS for Strategic Advantage

There is a downside to the strategic use of IS technology. One may adopt a strategy that fails or one which does not produce a sustainable advantage. Furthermore, a narrow focus on strategic use of IS technology may blind one to other changes occurring in the industry.

Warner (1987) discusses the tendency in the manufacturing sector to apply computers, and illustrates how some organizations have gone awry by careless use of computer technology. For example, Warner cites the case of a manufacturing firm which developed a sophisticated computerized warehouse facility just as the industry changed to Just-in-time delivery. He concludes that information technologies should be considered only after improvements and system reorganization attempts have been explored. Though Warner limits his discussion to the manufacturing sector, it seems clear that his caveats apply more broadly.

In a survey of 100 businesses that outperformed their industries Ghemawat (1986) concludes that managers must be concerned about the possibility of 'contestable advantages.' He notes that the distinction between sustainable and contestable is not a simple matter of degree. He suggests that the opportunity for sustainable advantage is best when an industry is undergoing a dramatic change in technology, demand patterns or input availability.

We feel that the issue of competitive advantage has other risks. The risks are magnified by the fact that use of information technology for competitive advantage is currently a fad among academics and managers alike. The major downside we see is that identification of strategic opportunities and assessment of the attendant risks is currently more of an art than a science. Where one has insightful, experienced practitioners one has a fair chance of achieving some success. However, if a group of inexperienced managers with limited vision attempt to adopt OIS as a strategic tool the results will vary from mediocre to disastrous. Consequently, managers must seriously address the following two issues:

DO WE HAVE THE IN-HOUSE CAPABILITY TO ATTEMPT TO USE OIS STRATEGICALLY?

CAN WE AFFORD A MISTAKE?

These issues focus managerial attention on the possible downside of using information technology as a competitive weapon. To use information technology competitively, managers must:

(1) focus on the core responsibilities of their organizations;
(2) identify key tasks and products and focus technology to support or enhance these areas;
(3) critically examine the dangers inherent in reliance on computer technology.

The dangers in category (3) include the potential for a major computer failure, the possibility of adopting an outmoded technology and the possibility that the technology will fail to deliver a usable product or a sustainable advantage.

Most of the instances where IS technology has been used successfully for strategic advantage are examples of the creation of a new product or service. Office information systems, as we will show, have a less direct strategic value.

STRATEGIC USE OF OIS

We maintain that OIS, as defined in Chapter 2, has a distinctly different role in organizational strategy than do MIS or plant information systems. The main differences are the integration and support roles played by OIS. By its very nature OIS leans toward a more thorough integration of activities than do MIS or other applications of information technology. Secondly, much of the application of OIS is in a support role rather than in a role which directly changes productivity. Consequently, OIS does not always lend itself to the generic strategies outlined for information systems.

OIS and Strategic Change

Office information systems can be thought of as having an indirect effect in terms of strategic advantage. However, there are a number of arguable direct effects. These include reduction of staff, enabling the organization to do things that it had not done before, and the ability to increase the levels of coordination. Most of these apparently direct effects are in fact indirect, since they support a particular strategy but do not constitute a strategy by themselves. We argue that the effect of OIS (as opposed to MIS which may be used strategically) is to create a new reality within the organization. The main consequence is that the organization has tremendously increased flexibility. Some might refer to this as degrees of freedom, sources of variation or the ability to create new alternatives. This potential flexibility itself creates a key issue for management. Some refer to it as the OIS challenge.

WHEN YOU CAN DO ANYTHING, HOW DO YOU FIGURE OUT WHAT YOU WANT TO DO?

Clearly, an important consideration is how to make the transition from a limiting, less flexible environment to a new state. Clearly this involves reassessing existing policies, goals and actions. However, it also requires that management assess the underlying models of their organization and of the organizational reality they have created. This is a very different kind of situation from consciously adopting a clearly defined strategy. In effect, the major issue for managers with OIS is:

HOW DO WE QUESTION OUR UNDERLYING MODELS OF THE ORGANIZATION AND HOW CAN WE DEVELOP NEW OR MORE EFFECTIVE ONES?

In addition they must help employees deal with the potential freedom associated with these new systems, which means that the employees' models of reality or models of appropriate behavior or organizational functioning, etc. will be subject to change. This may be very threatening to all employees. As one manager pointed out, the most common response to tremendous flexibility is total paralysis. The individual employee will look at a wide range of options and find it very difficult to make a choice, since he or she is totally overwhelmed by the variety of options.

While OIS is not a direct strategic weapon, it has a substantive indirect effect which changes the nature of the organizational context in which activities are done. Consequently, it enables the organization to conceive of new strategic directions. The limiting factor here is creativity, intelligence, and the amount of time available to use these abilities.

We have established that OIS has an indirect strategic value. But even as an indirect strategic weapon, OIS potential must be assessed in some way. Clearly, an organizational scan (as discussed in Chapter 2), looking for areas where there are problem spots and then asking questions regarding how OIS might help solve these would be one of the first steps. Another, more creative, step would be to instigate a meaningful wide-ranging dialogue with employees about these problem areas and about the way they do things now, to see if more creative alternatives could be developed.

A common approach is to introduce OIS, giving people an electronic version of what they have now. This gives them a chance to experience the system for a while. One can then promote a dialog about new ways to organize work processes. In this case people are arguing from experience with the technology, and how it affects their work. On the basis of their learning with the system they will be likely to identify new and more creative alternatives. Seen in this light, buying into an OIS implies that one is not buying a system, one is buying the first of a series of systems which will evolve as the corporation develops and changes in response to changes in the external and internal environments. Clearly, one of management's key responsibilities will be to manage ongoing change in the organization and ensure that a continuous process of assessment and reassessment is in place and is working properly. A further responsibility is to ensure that an appropriate electronic architecture exists upon which to build future systems. Keen (1988) notes that for the senior executive the telecommunications architecture is the technical strategy.

Some final thoughts are in order. With regard to strategic advantage, one of the key issues is the sustainability of an advantage. However, most of the writings about sustainability of competitive advantage tend to be a simple-minded argument concerning whether an advantage is sustainable or not. This is an inappropriate formulation of the problem. The appropriate question is to ask for how long an advantage is sustainable. In the long run no advantage is sustainable, because eventually other people will be able to duplicate what you do.

There are two aspects of sustainability that are worthy of attention. The first is the difficulty someone would face in trying to learn about and adopt the innovation. On the face of it, this is simple. If you can keep your innovation secret it is going to be harder for other people to copy it; although someone may spontaneously come to the same conclusions as you have. Furthermore, even if people can learn about your innovation, it may be very difficult for them to copy it because of high cost or high level of technical skills required. Second, it may be that in some particular areas your company is able to sustain a flow of enhancement to your strategic innovation such that you can maintain a position a step or two ahead of the competition for a considerable period of time. These are the traditional views of strategic advantage that were discussed previously.

There is a different way of looking at strategic alternatives. When considering strategic alternatives, ask the following question:

DOES THIS CHANGE WHAT WE DO, OR DOES IT CHANGE HOW WE THINK ABOUT WHAT WE DO?

If it merely changes what you do, then it is easy to copy. It may be expensive and may be time-consuming, but it can be copied. On the other hand, your innovation or your strategic thrust may change how people think about themselves, their corporation, their jobs, and their customers and enhance your organization's effectiveness, ability, attractiveness to customers, etc. In this case it is difficult to copy, because it involves changing the employees' conceptual models of the organization. One has only to remember numerous stories of corporate turnarounds to realize that many successful executives have adopted this strategy (e.g. changing the way people think).

This brings us full circle to OIS. Because of their inherent flexibility they break down previously existing barriers to communication, transfer of information and sharing of knowledge; they change a 'hardware environment' to a 'software environment,' or 'virtual environment.' This means that the potential for changing not only behavior, but the context in which behavior is modeled and perceived, is enormous. The challenge is to manage this potential effectively.

Applications of OIS

In this section we discuss the unique aspects of the strategic use of OIS. These include support for personal networking, support for unstructured activities and augmentation of key tasks.

Support for personal networking In Chapter 3 we discussed the research by Kotter, which showed how effective general managers spend their time. One important activity was personal networking—maintaining contacts with a wide range of people in the organization. OIS technology can play a major role in this activity by providing a variety of modes whereby a manager can remain in contact with the broadest possible range of others. The telephone system is one major contact medium. By extending its use through voice-mail facilities and conference calling a manager can expand his or her personal network. Computer mail and messaging systems, as well as computer conferencing systems, further expand the communication network. Being able to contact a wide range of people is not enough. Managers must be able to control the increased information load that the expanded communication network entails. Consequently, they require support for managing their communications.

Support for unstructured activities Much of the support for unstructured activities and communications is contained in current desktop software. Such features as customized mailing lists and telephone directories, as well as time management software, allow a manager to better control his or her communications. Furthermore, simplified databases allow a manager to keep records on each person contacted, their likes and dislikes and important dates such as birthdays, anniversaries, etc. This software can be used by individuals to manage their personal relationships within a business.

Augmentation of key tasks The use of spreadsheets, Query languages, graphics systems for data analysis and presentations, facilitates managerial work by augmenting key tasks. Furthermore, in the future more use will be made of public information networks (Videotext) to facilitate communication with the outside world. The key to augmenting the 'right' tasks is identifying key tasks in the first place.

MANAGEMENT ISSUES AND OIS

The key management issues related to the strategic use of OIS center around the extension of personal networking capability, the managing of personal information and the augmentation of key managerial tasks. The issues can be summarized as follows:

WHICH EXECUTIVES MOST NEED COMMUNICATION SUPPORT AND WHAT TYPE OF SUPPORT IS APPROPRIATE FOR EACH?

HOW CAN OIS ASSIST MANAGERS AND PROFESSIONALS IN MANAGING THEIR COMMUNICATIONS AND INFORMATION MORE EFFECTIVELY?

FOR EACH EXECUTIVE AND PROFESSIONAL, WHICH TASKS CAN BENEFIT MOST FROM AUGMENTATION, AND WHAT IS THE APPROPRIATE TYPE OF AUGMENTATION FOR EACH?

These three issues assume that managers are willing to examine their own activities and to adopt technology appropriate to their requirements.

This chapter has highlighted the issues that must be considered by management in the strategic use of OIS. The questions raised in this chapter are summarized in Table 4.1. It is important for management to attempt to answer each of these questions when considering use of OIS so investment dollars are not wasted on unproductive use or through unsuccessful implementation. In Chapter 7 we present a methodology which assists in addressing the issues raised here.

TABLE 4.1 Summary of management issues

---

(1) Which mix of generic strategies best fits our industry, our way of doing business, and our client base?
(2) Do we have the in-house capability to attempt to use OIS strategically?
(3) Can we afford a mistake?
(4) When you can do anything, how do you figure out what you want to do?
(5) How do we question our underlying models of the organization and how can we develop new or more effective ones?
(6) Does this change what we do, or does it change how we think about what we do?
(7) Which executives most need communication support and what type of support is appropriate for each?
(8) How can OIS assist managers and professionals in managing their communications and information more effectively?
(9) For each executive and professional, which tasks can benefit most from augmentation, and what is the appropriate type of augmentation for each?

BIBLIOGRAPHY

Curley, K. F. (1984) 'Are there any real benefits from office automation?', *Business Horizons*, July–August, pp. 37–42.

Frohman, A. H. (1982) 'Technology as a competitive weapon', *Harvard Business Review*, January–February, pp. 97–104.

Ghemawat, P. (1986) 'Sustainable Advantage', *Harvard Business Review*, September–October, pp. 53–58.

Grindlay, A. (1983) 'The computer as a competitive weapon', *Business Quarterly*, Summer, pp. 14–17.

Huber, G. P. (1984) 'The nature and design of post-industrial organizations', *Management Science*, **30**(8), 928–951.

Huber, G. P. and McDaniel, R. R. (1986) 'The decision-making paradigm of organizational design', *Management Science*, **33**(5), 572–589.

Kalbacker, W. (1983) 'Attacking new markets', *Computer Decisions*, 15 September, pp. 42–53.

Keen, P. G. W. (1988) *Competing in Time*, Ballanger, Cambridge, MA, p. 24.

Kleim, R. L. (1985) 'Does automation necessarily mean an increase in office productivity?', *Journal of System Management*, May, pp. 32–34.

Lasden, M. (1984) 'Playing technological catch-up', *Computer Decisions*, January, pp. 98–112.

McFarlan, F. W. (1984) 'Information technology changes the way you compete', *Harvard Business Review*, May–June, pp. 98–103.

Porter, M. E. (1985a) *Competitive Advantage*, Free Press, New York.

Porter, M. E. (1985b) 'Technology and competitive advantage', *Journal of Business Strategy*, Winter, pp. 60–68.

Porter, M. E. and Millar, V. E. (1985) 'How information gives you a competitive advantage', *Harvard Business Review*, July–August, pp. 149–160.

Roman, D. (1983) 'Information resources: the new competitive weapon', *Computer Decisions*, 15 September, pp. 30–38.

Strassman, P. A. (1985) 'The real cost of office automation', *Datamation*, 1 February, pp. 82–89.

Warner, T. N. (1987) 'Information technology as a competitive burden', *Sloan Management Review*, **29**(1), 55–61.

Wiseman, C. and MacMillan, I. C. (1984) 'Creating competitive weapons from information systems', *Journal of Business Strategy*, Fall, pp. 42–49.

Chapter 5
CONSEQUENCES OF OIS

INTRODUCTION

In this chapter we discuss the consequences of using OIS. While the implications of using these systems are not yet fully understood, it is important that managers identify the potential impacts on their organizations and on society at large. They must be concerned with the impact that changes in the social infrastructure will have on their ability to obtain and manage workers, and on their ability to manage their organizations. To address these issues we first discuss the societal impacts of computerization and then discuss the organizational impacts.

SOCIETAL IMPACTS

A major societal impact of OIS is a growth in jobs requiring computer expertise or knowledge, and the decline or attrition of jobs that require manual skill or effort. There is a further decline in those mid-management jobs which consist largely of information summary and transfer. This trend will accelerate as OIS absorb much of the work that was formerly done by human beings. As a consequence of these trends there are five social issues of particular interest to managers; they are job displacement, job security, job mobility, data security and individual privacy.

Job Displacement

It is clear that certain jobs are being displaced while others are being created. For example, three word processors can do the work of four typists; in some situations where there is a lot of editing (e.g. a law office) they may do the work of more. A study of the banking industry projected the displacement of two-thirds of all bank jobs with increasing office automation. Whether or not that prediction holds true, it is clear that many jobs can be replaced by office technology.

Many jobs have already been affected. Today, instead of having bank clerks record deposits and withdrawals, many of these transactions are done solely through computers. Similarly, in the insurance industry one finds that the number of claims processed are increasing as computerized systems allow a claims processor to handle more and more claims. Even in the police force one finds that automated systems are reducing much of the manual labor and therefore reducing some of the need for clerical workers.

As this trend continues we will find fewer and fewer clerical workers at the lowest level. It seems clear that we will also find fewer and fewer managers. In a recent article with the alarmist title 'Middle managers face extinction' (*Economist*, 1988), Peter Drucker was quoted as claiming that, with increasing use of information technology, whole ranks of middle management who act as information filters and communications pipelines will be phased out. He quoted statistics from companies such as Hanson Trust, which fired 250 head office staff when it took over Imperial. Similar stories abound. Presumably this 'trickle-up effect' will result in a leaner organization with fewer workers and fewer levels of management. The article concludes that the days of the human relay teams where middle management acts as 'human boosters' for the communication signals coming from the shop floor, are numbered.

If one type of job is being displaced and another type is created there will be some short-term or transitional problems affecting the job market. The fundamental societal issue that we face in the short term is:

CAN DISPLACED WORKERS BE MOVED TO NEW AND MORE APPROPRIATE OCCUPATIONS?

Given that taxpayers will have to pay much of the cost of supporting people who cannot find work, it is important for all taxpayers (individuals and organizations) to find out how fast new jobs are being created, how long some of these jobs will be sustained and what kind of education is required to help prepare a new generation of workers for a changed work environment.

At the same time society must retrain existing workers to meet the challenges of a rapidly changing workplace. An equitable arrangement for retraining workers is a necessity. A number of companies have their own retraining programs for workers who are displaced by office automation. These programs are commendable. In fact establishing such a program is very much in the organization's self-interest since it alleviates in some measure, the fear, mistrust and general concern often associated with office automation projects.

Who is to bear the costs of retraining for those workers who have no access to organizational retraining programs? We encourage a sharing of the responsibilities between the industries who profit from use of technology and government who must bear the cost of unemployment. Our argument is a simple one. We believe that it is economically and socially most efficient for organizations or industries to provide retraining programs; however, we recognize that this may not always be possible. Consequently, it seems that an industry-sponsored agency funded by tax incentives would be an ideal and economically efficient vehicle for providing retraining to displaced workers. In any event, society and industry must provide for those who suffer economic hardship as a result of technological innovation.

Job Security

Until approximately 20 years ago a worker, whether at a blue-collar or white-collar level, would be trained for a job and could reasonably expect to work at that job or profession for the rest of his or her life. Today, with the rapidly changing nature of the workplace as a consequence of computerization and other technological developments, we have gone from a concept of lifelong occupation to one of lifelong training. Unfortunately, however, our educational institutions have not kept up with this rather dramatic change.

In order for individuals to cope with the new social demands being placed on them, a two-pronged approach is needed. First we need a change in attitude. This change in attitude is as great as that which the government is currently trying to bring about with respect to drinking and driving. People must realize that a degree or a certificate or a diploma is no longer a lifetime guarantee of employment. We will have to begin to view our official training in university or community college as a first step in a lifelong process of training and re-education.

Individuals entering the job market should assume that they will have a succession of careers that may run the gamut from highly technical professional careers to more human-oriented careers. Even those who stay in a technical profession may change the nature of that technical position several times before they end their work life.

Changing individual attitudes is not enough; we must also change the nature of educational institutions. Most institutions of higher learning are designed to train someone for a career. It is clear that over the next five to ten years there must be a shift in organizational thinking. We shall still train people for careers but we shall have to do much more as well.

Career updating is one option. There is some recognition of this in business schools right now, where MBA updating programs are being developed. A second option involves programs designed for people who are reorienting their careers. For example, supposing someone with a highly technical degree wishes to change from one technical specialty to another related specialty. It is unlikely that a person will do a bachelor's degree in chemical engineering if he/she already has a degree in electrical engineering. However, right now there are few if any effective programs, other than masters programs, that are available to such people. A lot more thought must be given to the kinds of courses and upgrading that can teach people the skills required to switch professions effectively.

The implications for management are profound. Managers are going to need more flexible workers, and they are going to need upgrading of their existing workers. If these courses and skills are not currently being taught, organizations may have to develop their own training programs; otherwise, they will endure the problems associated with a workforce that is inappropriately trained for the new job requirements. The major societal issue is:

WHAT ARE THE IMPLICATIONS FOR INDIVIDUAL CAREER PATHS IN A SOCIETY WHERE JOBS AND PROFESSIONS ARE RAPIDLY CHANGING

Job Mobility

In a world where not only are jobs changing, but where the ability to shift jobs easily is going to become an advantage, individuals who have a sound generalist education coupled with specialized training in a particular area will have an advantage over those who have a very deep and extremely narrow education, or those who have very little skill. Consequently, society will have an unequal distribution of opportunities for employment. This will be coupled with a decrease in the number of unskilled jobs and a rapidly changing environment of highly skilled jobs.

One issue that both businesses and society must deal with is career paths for people in this dynamic environment. Most managers and many trained professionals are flexible and can develop their own career path over time. However, at the low end of the spectrum, where people have a lower level of education or skills, care is required to ensure that we are not training people through our community colleges and through industrial training programs and technical high schools, for dead-end work.

Based on the preceding discussion it seems likely that a sound generalist education followed by specialist training is probably the best basis for a career which will be highly mobile and involve a lot of shifts and re-orientations.

What would such training involve? It is clear that most jobs will, in the future, involve some work with computers. While it is unlikely that many professionals will be computer programmers, they will need to understand the technology as end-users.

Secondly, as managers do more and more of their own written output, and as professionals are relieved from the burden of some of the routine work, it is clear that writing skills and the ability to read quickly will be important job requirements in the future. Finally, as our analytic software improves, most employees must be able to interpret the output.

Consequently, the proper training for maximum job mobility will involve training in the use and understanding of computer systems; training in the ability to write well and quickly and to read large amounts of written information and digest it quickly; and finally, training in fundamental analytical skills such as the interpretation of statistical data.

This education would provide a good background for almost any job. Lower-level employees, who are trained to operate particular software, will require continual retraining as more and better software is made available. For example, the secretary who learns a word processing program will have to be prepared to continually learn new and advanced software programs if she wishes to remain mobile in the job market. Already we see examples of computer programmers who know only one language, that runs on some outdated hardware, and who are locked into their jobs for the rest of their lives because nobody else programs in that language. We will see more and more of this as time goes on. The key societal issue here is:

WHAT METHODS ARE BEST FOR DEVELOPING AND CONTINUALLY UPGRADING SKILLS IN THE POST-INDUSTRIAL SOCIETY?

Data Security

Data security is the ability to control access to information, and the maintenance of that information in an accessible and accurate state.

As more information is stored on computers, the ability of interested parties to tap into that information is going to increase. Whether it is government information which must be protected from unfriendly powers, or an organization requiring protection against industrial espionage, computerization increases the probability of an information leak. As more of this information becomes computerized it is increasingly subject to loss of security.

Most writers in the area would agree that no computer system is inherently secure. One writer has commented that the only completely secure computer system is a computer with no terminal that is locked in a windowless lead-lined room.

One important societal change must be the development of appropriate legislation to govern the acquisition and use of computer data. This legislative effort will require informed input from business as well as from human-rights activists. In addition there must be improved legislation to protect intellectual property stored in digital form. The societal issue here may be formulated as follows:

WHAT STEPS CAN SOCIETY TAKE TO ENSURE THE SECURITY OF IMPORTANT OR SENSITIVE DATA?

Privacy

The issue of privacy is controversial. If we start with the assumption that individuals have the right to control who has access to information about them, then it is clear that OIS poses a threat to individual privacy.

Anyone who has gone to a trade show and filled out a card with his or her name and address on it, or has given out a business card, will know, by the dramatic increase in junk mail, that good use has been made of that information. While junk mail is an annoyance, it is a minor example of the kind of violation of personal privacy that has resulted from extensive use of computers. Most credit card companies, many service organizations, and virtually all businesses develop extensive mailing lists of their clientele. These mailing lists are sold and resold to a variety of companies.

The advent of computers in the workplace has increased the potential for private information to be made public, or to be distributed to undesirable sources. One issue that unions are dealing with is the right of workers to privacy about their own performance. This issue is an attempt on the part of unions to limit the amount of information about individual performance that management can collect. This applies particularly to computerized information systems which have the ability to collect a variety of quantitative information about an individual performance for all employees, and to provide summaries to management so that employees could potentially be rated in terms of some quantitative output.

Other privacy issues are concerned with such things as one's address, phone number, income, family status, etc. One example of the widespread, and perhaps inappropriate, use of personal information, comes from the Province of Quebec. In Quebec they have a tax known as a water tax, which applies to every residential unit in Quebec. When moving from one living unit to another, anywhere in the Province of Quebec, it is necessary to fill out a form which gives details of the names of people who live in the house, their ages, income, professions, addresses at work or school, etc. This information is then collected at a municipal level and is thus not subject to provincial or federal information guidelines. Consequently, the information is shared among all the police forces in Quebec and among the municipalities, and is used as a way of tracking potential criminal elements. In fact the Quebec government is developing a computer system for the municipal police forces which specifically allows use of this information. While the threat to individual privacy may not be great, it is clear that the potential exists for this information to be used in an Orwellian manner.

Other threats to privacy come as a consequence of the large volumes of data that are collected. Another police example concerns the large integrated databases used both in Canada and in the US to record known or wanted criminals. The arrest some time ago of a stewardess in Atlanta, who happened to have the same name as someone for whom a warrant was issued in Texas, provides a rather shocking example of what can happen when erroneous data are fed into these databases. The woman was arrested coming off of an international flight where she had worked as a stewardess, and spent the weekend in jail before it was realized that she was not the person to whom the warrant pertained.

As databases grow larger and larger in the police sector, the credit sector and other areas of the economy, one can envisage a situation where, either as a result of malicious intent or merely through bureaucratic error, an individual's privacy could be severely violated.

There are no easy solutions either to security issues or privacy issues. Not only must managers be concerned, politicians and legislators must be involved as well. Businesses must be concerned about security of information and should be actively concerned about the legislation regarding how information is used and other potential impacts of large government and organizational databases.

Capture of certain kinds of information may generate large amounts of social debate even if the information is used solely for business purposes. An example would be the proposed development of a tenants registry by the Association of Landlords in Ontario. This Association proposed that they develop a registry of all people who rent from them, indicating whether or not they have ever had problems with these people in the past, such as not paying rent, damaging property, etc. On the surface this makes sense, and there are good business reasons for it. It is clear that if I as a property-owner wish to rent to someone, I would like to be able to identify in advance whether that person is unlikely to pay. However, the potential for abuse is equally high. For instance, if a landlord wished to seek revenge on a tenant over a dispute, all he/she would have to do is enter some negative information about the tenant into the file. This action would make it very difficult for that person to rent accommodation anywhere in Ontario.

Society has not yet developed appropriate measures for controlling large databases, or the acquisition and updating of information which goes into them. This situation is one which will not be easily resolved but which must be addressed over the next five to ten years. The privacy issue can be stated as:

WHAT MECHANISM CAN SOCIETY USE TO ENSURE INDIVIDUAL PRIVACY ON ONE HAND AND AT THE SAME TIME ALLOW EFFECTIVE USE OF INTEGRATIVE COMPUTER TECHNOLOGY?

ORGANIZATIONAL IMPACTS

While the societal impacts we have been discussing above clearly impact organizations, there are a number of consequences of OIS which directly affect the organization.

The idea that technology changes organizations is not new. In fact, for many years it has been quite clear that technology not only changes the structure and nature of organizations, but that it also changes the way people do their jobs, and to some extent the way individuals in organizations view themselves in the context of society. Some background will put these phenomena in perspective.

An Historical Perspective

Invention of high-speed steel, at the turn of the century, gave rise to the modern production shop and to mass production. Use of high-speed steel meant that machine tools could operate much faster than before. The potential for increasing the rate of production led Frederick Taylor to conduct his first studies on improving the organization of the shop floor. These studies eventually led to the theory of scientific management. This illustrates how a technical innovation (high-speed steel) led to the development of a particular kind of organization—the mechanical production shop. This led in turn to production lines, to the modern problems associated with them and to a certain philosophical view of man. Taylorism has imbedded in it the concept that men are essentially machines that can be manipulated to increase performance. If men are tools to be manipulated, it is appropriate to devise more efficient ways of manipulating them to get the highest possible productivity. This view has led to a number of economic successes and a number of social failures.

Technology not only changes the speed at which things operate, it changes the practical size of organizations and the way they are managed. Before the invention of the steamship, organizations such as the Hudson's Bay Company or the East India Company tended to give comprehensive local discretion to their Factors who were located in various far-flung sites. Part of this was due to the fact that the information available at head office in London could in no way be considered up-to-date. Consequently, senior executives had to rely heavily on the capabilities of their people in the field. However, with the development of the steam engine and the telegraph, communication was accelerated. Remote managers in London were able to exercise much more central control. The power of the individual representatives in the various countries was reduced. Since head office managers now had better and faster information, and could exercise more ongoing control, they changed from general instructions to their agents to more specific instructions about what to buy, and prices they attempt to negotiate, etc. In fact, technology decreased the psychological distance between the various remote locations and the head offices. The telephone has further accelerated this process by providing a means of communication anywhere in the world.

When computers were first introduced they were extremely large and unwieldy, though they calculated very quickly. Consequently the computer tended to cause a certain amount of centralization of information because very large, highly standardized programs were written for machines, at considerable expense. With the advent in the late 1970s of inexpensive personal computing this centralizing trend was reversed, because it became feasible to have decentralized computing at reasonably low cost.

Consequently, organizations are now starting to refragment. Information and data processing are being redistributed across all levels in the organization.

At the same time senior management has the ability to exercise greater control than before. These effects combine to push decision making closer to the bottom of the organizational hierarchy. The consequence of this has been to reduce the need for middle management. The article in the *Economist*, quoted earlier, gives the example of SAS (Scandinavian Airlines) which has inverted its management controls to give front-line workers the power to bypass managers and thus serve the client more effectively (*Economist*, 1988).

Distributing data capture, data storage, data processing and communications capability to every element in the organization tremendously increases the organization's degrees of freedom or requisite variety (Beer, 1974) available to the organization. This immediately raises questions of organizational structure and design.

Information Technology and Organizational Structure

An organization can only be structured to the extent that people have the ability to communicate with each other and to ship information and physical objects back and forth. When the only means of communication are written instructions via uncertain delivery systems, the nature of the organization will be different from when one has the ability for both voice and visual interaction.

Not only is the structure of organization changed, but work flows and communication patterns also change, because they are both a consequence of the way the organization is structured, and a determinant of it. Jobs which previously involved a significant amount of communication may involve less. Those who had little communication previously may have more.

The management issue from Chapter 2 can be restated as follows:

HOW SHOULD THE ORGANIZATION REDISTRIBUTE ITS ACTIVITIES TO TAKE ADVANTAGE OF THE FLEXIBILITY RESULTING FROM COMPUTERIZATION?

This restatement of the issue highlights the use of computerization as a tool for organizational redesign. However, when organizations face the variety of choices opened by computerization they typically try to replicate the past. This may be due to a lack of time to consider alternatives; it may be due to a lack of experience or to tunnel vision; and it may be due to the fear of change. Computers are a change agent and affect all areas of organizational life.

Organizational change can be viewed at both the macro level and the micro level. The macro view of computerization focuses on how computers can operate to change the organizational structure as well as the corporate culture. From a micro view, computers change both the communications that go on in an organization and the way in which employees do their jobs.

The Macro Perspective

The advent of a highly integrated, computerized office mandates that an extensive planning process, such as that described in Chapter 6, must occur if any benefits are to be realized. In addition to the planning *per se* there are a number of design issues which have implications for the organization as a whole. These include the following:

(1) The degree of centralization/decentralization of computer hardware, software, support, and administration that is appropriate.
(2) The nature of possible changes to corporate culture and the development of subcultures.
(3) The nature of fundamental changes to the way business is conducted.

There are a number of approaches to the issue of centralizing/decentralizing computer systems. Many organizations opt for centralized corporate databases for personnel, financial and customer data. Often distributed databases contain only data that are used within a particular work group or department. The rules by which one decides on the degree of centralized versus distributed processing and databases are not yet well established. It is a major decision and requires a lot of thought in the context of any particular organization. Furthermore, the degree to which computer support services and administration are decentralized are important *organizational design issues*.

Computers may also change the corporate culture. One potential change in corporate culture is the development of elite groups and particular subcultures focused around technical knowledge. They may develop a subculture which may or may not reflect organizational goals. The degree of resistance or support for implementing computer systems is partly a function of the norms of existing corporate subcultures. The success of an implementation of OIS may be substantially affected by the attitudes of these subcultures.

Computers can change the way business operates. They speed up some business processes and they can increase the focus on one or two elements of the business to the detriment of others. For example, as will be demonstrated in Chapter 9, computerized systems for the online monitoring of employee performance create profound changes in the nature of a business. First they bias performance toward quantifiable dimensions such as volume of customers, and away from qualitative dimensions such as quality of service. Second, they create interpersonal tension and conflict between workers. Finally, these systems can act as a barrier between front-line workers and their immediate supervisors. Taken together these changes can have a profound effect on the manner in which business is conducted.

The Micro Perspective

From a micro perspective we consider reorganization of jobs and redefinition of job requirements. One aspect is the redefinition of skill sets for a job. For example, the ability to use a word processor is now a part of job requirements for many secretarial jobs. Changes can be widespread. As managers tend to use word processing more they will need secretaries with editorial and copy-editing skills. A second way in which job requirements change is through the repackaging of the tasks which comprise the job. This could involve adding new elements to the job such as improved customer service. It may involve completely repackaging the task to make a more comprehensive controllable job from one that had been rather fragmented.

An example involves checkout clerks in grocery stores. The advent of point of sale (POS) terminals to replace the old cash registers has caused a tremendous emphasis on productivity. This has affected the quality of interaction with the customers.

Office information systems also affect the nature of the communications. They affect the quality, and perhaps the frequency, of communication. As people become more oriented toward terminals, communications may tend to be more impersonal through the use of electronic mail. However, people working on terminals all day may feel that they can communicate over time and distance more reliably. For example, many systems now have the ability to allow the caller to be notified when the receiver has viewed the message.

Job design and redesign is critical, since changing the nature of a job not only changes what an individual does, it changes the links between workers and the links between workgroups. In fact, information and products may take a different path throughout the organization than they did before. OIS also changes the emphasis placed on different aspects of task performance. For example, if one were to redesign office jobs so as to create a production line environment in the office there would be a shift in emphasis towards volume as a basis of performance and a shift away from the more interpersonal and custom-oriented aspects of the job.

Given that OIS change the communication links between the workers, and given that many of the systems tend to increase the emphasis on productivity, many individuals find an increase in their levels of stress. This increase in stress can cause a number of health-related problems, a loss of interest in the job, feelings of powerlessness and a variety of other physical problems. A number of studies have shown that most tired eyes, aching backs, insomnia, character changes, feelings of social isolation and deep depression attributed to computers, can be attributed to the poor planning and design of these systems. Many of the problems are a result of a lack of basic human needs.

As an example of this, some offices limit the ability of clerical workers to talk to each other while doing their job, time their washroom breaks, and control the order and variety that they find in their work. The consequences of this highly structured and alienating environment are profound and undesirable. Most writers in the area of OIS conclude that the undesirable social impacts of the system are the consequence of the characteristics of the design and implementation of the system, and are not inherent in the computerized systems themselves.

Technology also affects the security of information. It provides yet another path for unethical employees to penetrate sensitive information. At the organizational level there are a number of actions one can take to increase security. A first step is to provide a clear system of authorization for access to different levels of information. The second step one can take is to design systems which allow everyone access to information but which make it extremely difficult for anyone to copy information. This design would produce a distributed system with centralized file servers to which diskless workstations were attached. While an employee could always obtain and manipulate data, he or she could not pull off a diskette containing critical financial data for the corporation and sell it to a competitor. Even this provides limited security.

Disaster prevention is another aspect of security. In the event of some catastrophic failure the organization needs some plan to ensure that vital data are not lost. Currently there are a number of companies who provide vaults in which one can store tapes of the day's transactions against the potentially catastrophic computer failure. At worst one would lose only a day's worth of transactions.

Management's role in the design of catastrophe backup systems is three-fold. First, management must ensure that the OIS people have provided a satisfactory catastrophe protection program. Second, management should review the program with all parties concerned and ensure they are aware of their roles. Third, the plan should be reviewed by the EDP auditors for sufficient levels of planning and control.

At the micro level we must also consider physical, or ergonomic, issues that are associated with use of the terminals. It is important for managers to realize that when dealing with computers as a change agent there is no one solution. The solution you arrive at will depend upon the conditions in your organization.

Perhaps the most fundamental impact is the redesign of work and workflow within the organization. It is important to note here that most OIS failures result not from technical inadequacies, but from social and organizational inadequacies in their design. In many organizations the need to consider the requirements of the individual are given at best lip-service. On the other hand, it is very difficult to anticipate all requirements for OIS and the resulting changes in organizational structure. It is not easy to conduct a relevant-needs analysis that provides managers with exactly the information they require.

Finally, OIS and computerization affect the quality of work life. Are we creating jobs which are more meaningful and challenging, or are we taking away much of the interesting work and giving people a tremendous amount of high-pressure, boring, routine work to do?

WHAT CAN MANAGERS DO?

Perhaps the most important thing managers can do is allow sufficient time for planning and consultation. They must be sure that the various workgroups that are going to be affected by the office information systems have an opportunity to provide input to the design of these systems.

Typically, this has not been done very well. An example of this is given in a study by Willem A. Wagenaar (1985). In that study Wagenaar compared VDT users to non-VDT users. In every case he found that the VDT users were more bored with their work, were more likely to be unable to choose their own work, were more likely to dislike their work load, were more apt to be behind in their work at least once a week, were more dissatisfied with the pace of their work, and were more worried about reprimands than non-VDT users. This study indicates that the way in which Wagenaar's sample of VDT users were treated in the office tended to cause rather more negative perceptions of their jobs than were found in a sample of non-VDT users.

It is important, given the diversity of society and organizational consequences, to review the issues that society and managers must face over the next five to ten years. Table 5.1 lists the issues that must be addressed.

TABLE 5.1   Societal and organizational issues

(1) Can displaced workers be moved to new and more appropriate occupations?
(2) What are the implications for individual career paths in a society where jobs and professions are rapidly changing?
(3) What methods are best for developing and continually upgrading skills in the post-industrial society?
(4) What steps can society take to ensure the security of important or sensitive data?
(5) What mechanism can society use to ensure individual privacy on one hand and at the same time allow effective use of integrative computer technology?
(6) How should the organization redistribute its activities to take advantage of the flexibility resulting from computerization?

Naturally, not all of these issues can be addressed by each manager; however, it is important for each manager to be aware of them.

Our goal is first to sensitize readers to possible consequences of OIS, and secondly to provide a process whereby each reader can construct his or her own solutions to key issues in the context of their own organizations. As an exercise in self-education, the reader may wish to apply these issues to Minicase 5.1, at the end of this chapter. While there are no answers in the back of the book, the process of thinking through the questions will provide useful insights.

In Part II, which follows, operational methods for planning, needs analysis and implementation are presented.

MINICASE 5.1: THE DENTAL CLAIMS SYSTEM

The processing of dental claims represents a major cost for most Canadian insurance companies. Typically these claims are received on a 'standard claim form' from an individual who is a member of a group dental plan.

The forms are sorted and placed in temporary storage. Each day armies of clerical workers enter data from the forms into their organization's claim system. The workers sit in rows in expensive, ergonomically correct chairs. Most work on a pay-for-performance system where volume of claims processed determines their take-home pay. The workers are typically women, and in large urban centers are typically recent immigrants. Not only are the working conditions restrictive (timed washroom breaks, etc.), ethnic and racial strife mar the work environment. In good economic times the turnover rate in this job is around 30%.

The insurance companies in southern Ontario are currently considering a joint system to automate the dental claims process. This system would result in an electronic hookup to each dentist's office. The dentist's office staff would fill out an electronic version of the standard claim form and electronically transmit it to a clearing center which could route it to the appropriate insurance company. While the system could be costly to implement, it has the potential to eliminate roughly 4000 jobs, and make the dental claims process faster, more accurate and less hassle for the subscriber.

This is clearly an instance where task automation is being considered. It raises the following questions:

(1) What would a system that was designed with an augmentation perspective look like?
(2) What should be done with the displaced workers?
(3) Whose responsibility is it anyway?
(4) What is the role (if any) of government in this?
(5) What privacy issues are raised by this system?

BIBLIOGRAPHY

Atkinson, P. (1985) 'People left out', *Office Equipment and Methods*, April, pp. 34–35.
Beer, S. (1974) *Designing freedom*, Canadian Broadcasting Corporation, Toronto.
Benbasat, I., Dexter, A. S. and Masulis, P. S. (1981) 'An experimental study of the human/computer interface', *Communications of the ACM*, **24**(11), 752–762.
Booth, P. J. (1982) *Study of the Social Impacts of Office Automation*, Wescom Communications, Ottawa.
Brown, D. (1984) 'Networks: a user's perspective', *CA Magazine*, **117**, 94–98.

Culnan, M. J. (1984) 'The dimensions of accessibility to online information: implications for implementing office information systems', *ACM Transactions on Office Information Systems*, **2**, 141–151.

Diebold, J. (1984) 'How new technologies are making the automated office more human', *Management Review*, **73**, 9–33.

Duffy, J. (1985) 'Communications will never be the same', *Office Equipment and Methods*, January–February, pp. 87–92.

Economist (1988) 'Middle managers face extinction', *Economist*, 23 January, p. 59.

Flynn, D. and Foster, L. (1984) 'Management information technology: its effects on organizational form and function', *MIS Quarterly*, December, pp. 229–235.

Hart, M. (1984) 'How the office of the future is shaping up', *CA Magazine*, **117**, 72–76.

Holley, C. and Reynolds, K. (1984) 'Audit concerns in an on-line distributed computer network', *Journal of Systems Management*, June, pp. 32–38.

Hussain, D. and Hussain, K. M. (1981) *Information Processing Systems for Management*, Richard D. Irwin, Homewood, IL.

Ivancevich, J. H., Napier, A. and Wetherbe, J. (1983) 'Occupational stress, attitudes and health problems in the information systems professional', *Communications of the ACM*, **26**(10), 800–806.

Keen, P. G. W. and Woodman, L. A. (1984) 'What to do with all those micros', *Harvard Business Review*, September–October, pp. 142–150.

Olson, M. H. (1983) 'Remote office work: changing work patterns in space and time', *Communications of the ACM*, **26**(3), 182–187.

Pickworth, J. R. (1983) *Productivity: Three Perspectives*, Productivity Research Group of Canada, Toronto.

Sanders, L. (1984) 'The impact of DSS on organizational communication', *Information and Management*, June, pp. 141–148.

Sauter, S. L., Gotlieb, M. S., Jones, K C., Dodson, V. N. and Rohrer, K. M. (1983) 'Job and health implications of VDT use: initial results of the Wisconsin–NIOSH study', *Communications of the ACM*, **26**(4), 284.

Strassmann, P. (1985) *Information Payoff*, Collier Macmillan, New York, USA.

Tapscott, D., Henderson, D. and Greenberg, M. (1985) *Planning for Integrated Office Systems*, Holt Rinehart & Winston, Toronto, Canada.

Thode, J. (1984) 'System security requires technical and user controls', *Journal of Systems Management*, June, pp. 38–40.

Vacca, J. (1983) 'Planning an information system', *Infosystems*, December, pp. 90–92.

Wagenaar, W. A. (1985) 'The psychological costs of master computer', *Datamation*, **31**(13), 157–162.

Part II
OPERATIONAL ISSUES IN OIS

The three chapters in this section present the issues managers must face when planning, implementing and evaluating OIS. The planning process discussed in Chapter 6 provides a structure in which an operational plan for an OIS may be constructed. Chapter 7 presents a methodology for user needs analysis. While needs analysis is properly part of the planning process, it is sufficiently important that it deserves to be considered separately. Implementation and evaluation are discussed in Chapter 8.

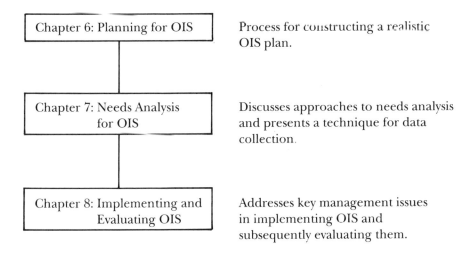

Chapter 6: Planning for OIS — Process for constructing a realistic OIS plan.

Chapter 7: Needs Analysis for OIS — Discusses approaches to needs analysis and presents a technique for data collection.

Chapter 8: Implementing and Evaluating OIS — Addresses key management issues in implementing OIS and subsequently evaluating them.

Chapter 6
PLANNING FOR OIS

There are as many reasons for not planning an OIS as there are organizations who lack a plan. However, as one moves toward an integrated office system, planning carefully becomes of greater importance. There are several reasons for this.

First, an integrated system requires that its components mesh smoothly together. Second, development of OIS is a costly, long-term activity that entails some risk. Finally, development of an appropriate OIS requires knowledge and expertise from many sectors and levels within the company. Each of these reasons is sufficient justification for a planning effort. Together, they mandate a serious planning process as a key to successful OIS design and implementation.

BENEFITS OF OIS PLANNING

The following benefits can be obtained by planning for the OIS (McFarlan, McKenney and Pyburn, 1983, list similar benefits for information systems):

(1) development of user confidence and acceptance;
(2) minimizing effects of technological turbulence;
(3) maximizing gains from technological opportunities;
(4) development of appropriate work flow analyses;
(5) meeting short-term needs in the context of long-term goals;
(6) effective use of corporate resources.

We discuss these benefits in detail in the following sections.

Developing Confidence and Acceptance

A well-planned implementation of office automation can provide the organization with sufficient time to encourage acceptance of the new system and to increase the confidence of both users and management in the benefits of OIS. Implementations where new technology was introduced too quickly are uniform in their lack of success (Ackoff, 1967; Carter and Silverman, 1980; Elton and Carey, 1981; Keen, 1988; Tapscott, 1982; Tapscott, Henderson and Greenberg, 1985).

Managers may feel that their power, authority and status are affected by these systems. To gain acceptance the implementation team must be willing to recognize these fears. A good description of this problem is given by Argyris

(1971). He argues that introducing information systems will cause resistance because executives will feel threatened. Though this article is almost 20 years old, the problems are still with us. In the early stages of an office automation planning exercise the OIS team must be careful to include senior managers. In the latter stages middle management must be involved, as must operational-level employees.

End-users may also feel threatened by OIS. Many people are afraid of losing their job or of learning a new skill. Many have much time and resources invested in current processes and will be unreceptive to changes in secure routines. Time spent by the planning team early in the process to answer questions and allay fears will have an immense payoff later on. Furthermore, user involvement begins their 'investment' in the new skills and job processes.

User involvement will not only reduce resistance, it can also enhance the final product. Those who actually do the work will likely provide valuable insights regarding changes in current work-handling processes. In our own research we have found that the front-line workers inevitably have useful insights on work redesign. Other researchers such as Enid Mumford (1981) go further. Using her ETHICS technique, Mumford has had clerical workers redesign their job functions and support technology.

Technological Turbulence

One of the current difficulties with the available technologies is that industry-wide standards have not yet been developed (see the discussion in Chapter 3). Consequently, users may require systems that cannot easily interact with other systems. Given the turbulent nature of today's industry, those who install OIS systems without a coherent plan are virtually assured of incurring unnecessary conversion costs at a later date. Much of the problem can be avoided by having an acquisition plan. Currently, the best rule is to trade off cost in favour of system flexibility.

Maximizing Gains from Technological Opportunities

The existence of a detailed office automation plan enables the maximization of gains from OIS technology in the following ways:

(1) negotiation of advantageous package deals from vendors;
(2) phased acquisition of enhancements and improvements to new and old systems in the context of a strategic OIS architecture;
(3) effective utilization of equipment and experience organization-wide.

While the first two points are reasonably self-explanatory, the last one needs some elaboration. Essentially it refers to the transfer of experience with various systems to different areas in the organization. A department undertaking a project may achieve a better return on investment through the involvement of other groups with similar problems.

Work Flow Analysis

Another aspect of planning is the collection of accurate data on the information and communication processing in each organizational unit. By developing both long- and short-term plans the work flow data required to implement a full-scale system can be obtained during the course of solving day-to-day problems. The organization will then be in a strong position to take advantage of new developments quickly, since much of the preliminary work flow analysis will have been completed.

Short-term Needs and Long-term Goals

Since no 'off-the-shelf' integrated office automation package exists, it is necessary to move in a step-by-step fashion. This approach fits the solution of today's problems in the context of tomorrow's requirements. Typically, immediate requirements are for communication systems or for text handling systems to solve specific problems. The dangers in dealing with these problems individually are:

(1) significant economies of scale can be lost;
(2) when a move is made to integration, significant extra conversion costs will be incurred;
(3) unnecessary duplication of hardware, software and training can occur.

With a comprehensive plan that allows the dovetailing of short-term decisions with long-term goals these pitfalls can be avoided. Additionally, decisions on expenditures can be planned rather than dealt with on an *ad-hoc* basis (as we mentioned in Chapter 2, the organizational scan provides limits within which costs can be planned).

Effective Resource Utilization

A well-planned implementation process allows senior management time to reallocate financial and human resources. By introducing integrated systems, management is redesigning the organization. Some jobs will be changed, some will be eliminated, new jobs will be created. Additionally, time is needed for management and employees to comprehend the new system and adjust to it. An appropriate planning process allows a time buffer within which this adjustment can be accommodated.

A well-designed OIS plan puts management in control. If the technology is not working as planned, or if the organizational effects are undesirable, management can stop the process. Since much of the office technology is inexpensive (personal computers, for example), it tends to proliferate if constraints are not instituted. Recently, a vice-president of a major financial institution was asked by one of the authors how many personal computers his organization owned. His first answer was ten. Upon checking, he found that the number was 200. While the unit cost is only $3000–$4000, the total cost of a personal computer is normally estimated to be in the order of $25 000 when software, associated hardware, training and maintenance contracts are included. The organization spent an uncontrolled $5 million. A thoughtful OIS plan could have ameliorated this.

A FRAMEWORK FOR IS PLANNING

Here we present some generally accepted concepts, together with a description of the special requirements of information systems planning.

General Planning Concepts

Several authors of planning articles (e.g. Bowman, 1983; McConkey, 1981; McLean and Soden, 1977; Oliver and Garber, 1983; Keen, 1988; Tapscott, Henderson and Greenberg, 1985) make the following (or similar) distinctions between different kinds of planning:

(1) *Strategic, long-range planning*: covers a five- to ten-year period and identifies goals and purposes but not the specifics of how to achieve these. In this context we discuss an IS strategic plan.

(2) *Medium-range planning*: deals with specific objectives that can be achieved in three to five years and identifies projects that are necessary to achieve these.

(3) *Short-range planning*: deals with a one- to two-year time frame and defines specific actions to be taken. Items in this plan will be justified and included in short-term budget requests.

The IS Strategic Plan

The strategic, long-range IS plan contains a statement of the goals and purposes that are to be achieved and describes future scenarios. It is an executive level plan and should include:

(1) Organizational goals to be achieved with IS (e.g. integrating telecommunications, office services and data processing into one unit and developing information resource management).

(2) Technical goals to be achieved with IS (e.g. developing a fully interactive network so that all intelligent devices have the capability to interact with each other).

(3) Functional goals to be achieved with IS (e.g. to provide sufficient support such that every manager and professional will have a workstation within ten years and be able to use it for at least 40% of his/her activities).

(4) Interfaces with other strategic plans in effect. It is particularly important to connect the IS strategic plan to other major planning efforts (Ghosh and Nee, 1983; Keen, 1988; Tapscott, Henderson and Greenberg, 1985).

Since IS has the potential to restructure large portions of the organization, it cannot be done in isolation. Kay *et al.* (1980) discuss this with respect to information systems. Keen (1988) raises similar issues with respect to planning a telecommunications architecture for the organization. The strategic plan should discuss the overall corporate scenario five to ten years in the future. The following questions help identify this future configuration:

(1) What will the structure look like?
(2) What will be the mix of managerial, professional and support staff?
(3) Will the organization as a whole grow or shrink in terms of physical size, number of staff and volume of business?

The plan should also describe what the IS system will include when complete. While this view may not reflect what is actually produced, the process of developing a detailed description will raise important issues and will provide a framework for discussion when the plan is finally presented. Some of the issues that must be addressed are:

(1) *The design philosophy*: top-down, bottom-up or middle-out. Top-down design begins with the goals of senior management and works down. Bottom-up design begins with specific activities and aggregates these upward. Middle-out starts at the middle management level and goes up and down the hierarchy. These approaches are discussed with reference to information system design by Ahituv and Neumann (1982, pp. 167–170).

(2) *The technologies included*: for example, mainframes, minicomputers, micro-computers, telephone switches, local area networks, etc.

(3) *The functionalities included*: for example, voice-mail, word processing, decision support systems, database management.

Finally, the strategic plan should identify a set of medium-term milestones which identify the path to the long-term goals. These milestones will be rather general and may change over time. The milestones serve as benchmarks against which the success of medium-term plans can be judged.

The strategic plan should be updated in a minor way every one to two years and should receive major rethinking every three to four years.

The Medium-range Plan

The medium-range plan is essentially a master plan covering three to five years. In this plan, specific projects are identified which, if accomplished, should lead to the achievement of milestones identified in the strategic plan. This plan should include:

(1) a statement of subgoals to be achieved (milestones);
(2) a description of proposed projects and priorities;
(3) a resource utilization plan (manpower, equipment and capital).

This plan is aimed at the managerial level and should be sufficiently detailed that managers in each area will know the resources required of them and the tradeoffs available to them.

The Short-term Plan

This one- to two-year plan is an operational plan which includes specifics of the projects undertaken, what is to be achieved by each and the resources required. This plan is the most detailed and specific of the three.

UNIQUE REQUIREMENTS OF OIS PLANNING

Due to technological turbulence, and the fact that identification of user requirements is still an art rather than a science, OIS planning has some unique characteristics. These are discussed in the sections which follow.

Involvement of Management

Earlier (page 77) we mentioned that OIS involves a process of organizational redesign. If OIS planning does not have support at the senior level it will have little success (see Barrett and Oman, 1983). Management provides guidance for other staff members' acceptance, and has the authority over critical resources required for the project's success. Consequently, the search for a senior corporate mentor is an important first step. If this support is not forthcoming, then comprehensive OIS planning is not possible.

Localized planning to implement OIS systems within a department or workgroup can still be instigated. However, one must also be careful to ensure the support of middle management (see Oliver and Garber, 1983).

In addition to enhancing activities in one area, a localized plan may serve as a demonstration project which will convince management to support more comprehensive OIS planning.

Involvement of Users

The involvement of users has received much discussion in information systems planning. It is also vital to the success of OIS planning (Tapscott, 1982, discusses this issue at some length). Office automation goes further than providing information. It challenges and potentially changes the way an individual does his or her job. To be successful an OIS implementation must have the active support of the majority of users. Otherwise, expensive workstations and sophisticated telecommunication systems will go unused. Users will become actively involved when:

(1) the new services meet a recognized need;
(2) they feel a commitment to use the new system.

The best way to ensure that these conditions are met is to have user representatives involved through the planning process. Bryce (1983) suggests that users should be made responsible for ensuring their needs are met.

Pilot Trials

Due to the potential impact of office automation and the existing uncertainties about individuals' reactions to various OIS technologies, it is wise to begin with pilot trials of the various subsystems in localized and highly receptive areas (Asner and King, 1981, make a similar suggestion). The advantages of this approach are two-fold: first, if the trial is unsuccessful the negative impact on future endeavours will be reduced; second, if the trial is successful a gradual and phased introduction of the technology to the rest of the organization can be adopted. This gives individuals time to adapt to the technology and changes in work habits.

A FRAMEWORK FOR OIS PLANNING

A four-step approach to office automation planning is presented in Figure 6.1. The first step is the formation of an OIS team which is responsible for developing the strategic, medium- and short-term plans. At the completion of each planning phase the plan is presented to management for approval. The outcome is a decision either to revise the plan, stop planning completely or to approve the plan and proceed to the next step.

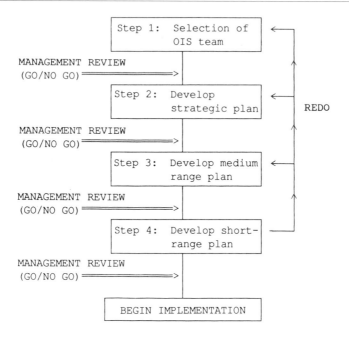

FIGURE 6.1   A Four-step approach to OIS planning

Selection of the OIS Team

It is essential that senior management establish a team of individuals from within the organization which is responsible for planning the office automation strategy and implementing various technologies. This team must include representatives from data processing, office services and telecommunications, as well as members from various user groups (management and support staff). Their involvement will provide the OIS program with a wide range of perspectives and expertise (Buss, 1983; Irving and Munro, 1983; Keen, 1988; Tapscott, 1982).

The OIS team should report to senior management, ensuring management's awareness of potential problems and benefits and obtaining their approval and support throughout the OIS project. Management should appoint a team leader to co-ordinate contacts regarding the office automation programme (e.g. consultants, vendors, users).

Reporting Requirements of OIS Team

The main reason for having the OIS team report to a senior-level manager is that the strategic planning phase will involve goal setting and the redesign of organizational structure and workflow systems. An additional reason is that this

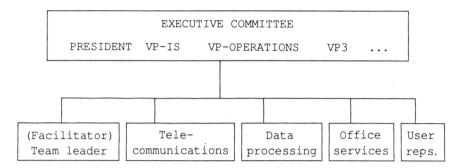

FIGURE 6.2   Reporting requirements of OIS team

reporting relationship will give added clout to the team. The overall reporting structure of the team is illustrated in Figure 6.2.

Purposes of OIS Team

The OIS team can serve several purposes including:

(1) Reducing internal power struggles for control of OIS and ensuring that an organization-wide (macro) approach is adopted.
(2) Ensuring development of communications throughout the organization regarding office automation, making staff aware of its objectives, keeping them informed of plans for change and gathering feedback from workers about office automation.
(3) Encouraging participation in office automation projects throughout the organization.
(4) Identifying competent people who can contribute to the office automation study.
(5) Evaluating existing systems and organizational procedures (there is no benefit in automating inefficient procedures).
(6) Ensuring that the capabilities of present equipment are utilized.
(7) Preventing unnecessary duplication of hardware or software within the organization.
(8) Assuming responsibility for the specification and implementation of new technology and the post-implementation evaluation of technology.

Operation and Management of OIS Team

Since the OIS team will be dealing with complex and controversial issues, it must be managed in an open and non-directive manner. The team leader should operate as a meeting chairman and should encourage expression of minority opinion. This does not mean that the team leader is inactive. He or she will take an active role in agenda setting, moderation of disputes and goal setting (those who would like more information on non-directive leadership are encouraged to read Maier and Maier, 1957; Maier and Solem, 1952).

While the non-directive approach will be slow and frustrating at first, it is essential for two reasons: first, all relevant issues must be presented and discussed if an appropriate plan is to be developed; second, debating a wide range of issues early on will provide a basis for dealing with issues raised by management and user committees at a later stage. (McKeen's (1983) review of MIS application development showed that successful projects tended to spend more time in analysis than did unsuccessful ones.)

Because of the complex and potentially controversial nature of the OIS planning process, the selection of the team leader is crucial to the performance of the OIS team. It is this person who must mitigate the frustrating process of the non-directive approach to bring about eventual group consensus. The main criteria for leader selection are an openness to a variety of opinions, credibility within the organization, sufficient technical knowledge to be able to interact with technical people, and good interpersonal skills. In many cases management will select an outside consultant to act as team leader or facilitator. While such a team leader will not know the organization well, there are often advantages to this strategy. An outsider usually has a broad perspective on office automation. In addition, he or she is usually not perceived to be biased toward certain departments. This credible impartiality can help the OIS team grapple with interdepartmental issues.

Selecting User Representatives

The involvement of user representatives is important to the success of office automation implementations. The level of the users involved will vary with the stage of the OIS planning and implementation process.

During the strategic planning phase, senior management must be involved. This will entail meetings with the executive committee, having a senior vice-president sit in on team meetings and having the director of planning or human resources involved throughout the strategic planning process. At this stage the OIS team is attempting to reconcile OIS with the strategic vision of the organization; and at the same time using the opportunities presented by OIS to broaden the strategic vision of senior management. This is the point at which the preliminary data from application of the HIT technique (see Chapter 7) are used. This provides a consensus-based overview of the organization which serves as a starting point for organizational redesign efforts.

At the level of medium-term plans, line managers whose departments are most affected will be involved. This is necessary to obtain their commitment and their operational insights which will be necessary for successful implementation. Finally, at the short-term level, operational personnel in a department must be involved, since they will be directly affected by the projects to be implemented. The relationship between these three levels of planning is illustrated in Figure 6.3.

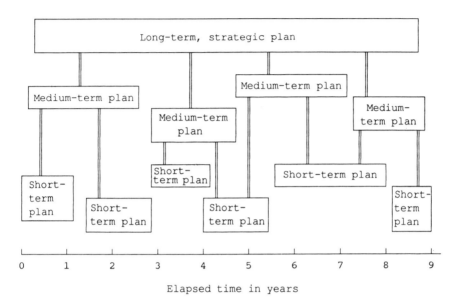

FIGURE 6.3 Relation between long-, medium- and short-term plans

Development of the Strategic, Long-term Plan

The development of a strategic plan is the first and, in many ways, the most important task of the OIS team. Since the OIS strategic plan must fit with other long-term plans the organization is pursuing, interaction with senior levels of management will be intensive. In order to develop a set of long-term goals and formulate a strategy for obtaining them, the OIS team must have a clear vision of what the organization will look like in five to ten years. This vision must come from senior management if it is to make sense in the long term. For example, management may plan to substantially change the formal organization over a five-year period and increase staff by 10–20%. If the OIS team develops a strategy based on the current structure and a constant staffing level, the results will be of little value. As we mentioned in the preceding paragraphs, data from initial needs assessments such as HIT are useful at this stage.

After obtaining or developing a set of long-term organizational goals the OIS team must assess the contribution OIS can make to goal achievement.

This could involve a number of approaches, among which is analysis of the value chain (Porter, 1985).

To assess an organization or a product using a value chain analysis one must discern different sequences in the organization's primary activities and determine which are overall organizational support activities. For example, OIS can help link the primary and support activities through a better communication system for file exchange, or through more timely information provided in spreadsheets versus having to calculate operational reports manually. This fast exchange of information could help an organization service its customers better or help determine competitive advantages.

While an organization-wide perspective is necessary at this stage, there should be sufficient attention to detail to indicate the generic technologies involved and to present a crude timetable for goal achievement. The completed strategic plan should include the following:

(1) an overview of the whole planning process;
(2) a statement of long-term organization goals for a five- to ten-year period;
(3) a description of the role OIS technology can play in achieving these goals;
(4) an analysis of likely requirements for successful implementation of OIS;
(5) a timetable for sub-goals in units of three to five years. These become the goals for successive medium-range plans (see Figure 6.3);
(6) global estimates of resource requirements over the five- to ten-year period;
(7) overall description of benefits to be obtained, the required risks and a discussion of how goal achievement is to be measured and costs justified.

Development of the Medium-range Plan

The medium-range plan, covering three to five years, will attempt to meet the various subgoals set by the strategic plan. In this case specific projects which contribute to overall goal achievement are identified for each functional area.

At this stage middle management will be heavily involved as members of the OIS team and as resource persons (Easterby-Smith and Davies, 1983 discuss the problem of developing strategic thinking among managers). The medium-range plan should include the following items:

(1) a needs assessment for OIS by organizational subgroup or by functional groups (e.g. marketing department, engineering department, purchasing, etc.);
(2) a technology assessment;
(3) a schedule of projects which relates needs and technological capabilities to the goals to be achieved.

These three factors are discussed below.

Needs Assessment

The assessment of needs or requirements for office automation in an organization can be a complex, specialized and costly task. The quality of the assessment depends on the level of expertise of those who conduct it, the time available, the level of cooperation of all participants and the financial resources allocated to it. Other contributing factors are the size of the organization and the scope of automation being considered. Subsequently, in Chapter 7, an operational methodology for needs assessment is presented. In this chapter we are more concerned with the process at a macro level. Five generic components of the needs assessment process are discussed. These are:

(1) analysis of existing goals, subgoals and procedures for goal attainment;
(2) analysis of use of existing communication information systems;
(3) identification of high-benefit areas for office automation; and
(4) production of an OIS opportunities report;
(5) a management review of the opportunities report and a decision as to the continuation of OIS.

Analysis of existing goals and procedures   In this step the OIS team attempts to define the main tasks for each organizational subgroup in terms of the organizational goals that pertain to it (see the Functional Analysis Form in Chapter 7). While the level of analysis will vary, it is unlikely to be useful to carry the analysis beyond the level of the workgroup (i.e. superior plus immediate subordinates). The main reason is that at the next lower level (the individual), one finds that data tend to be falsified as a result of the interpersonal threat. Furthermore, analysis at this level generates too much data to be of use to management.

A number of approaches are available to collect the required data. They range from interviews with a few key people to detailed analyses of individual tasks. We recommend focusing on the workgroup and the key tasks for which it is responsible. The resulting analysis should provide sufficient detail so that the OIS team can distinguish between those areas and functions which:

(1) need not be considered further;
(2) can be improved by reorganization alone;
(3) can be improved by OIS alone;
(4) can be improved by reorganization and OIS together.

Since this analysis has both policy and structural implications, the support and cooperation of senior management are essential to its success (see the Tasks and Links Report in Chapter 7).

Analysis of existing communication–information systems   This step involves conducting a survey of existing systems (e.g. telephone, copying, mail, data processing) to determine how they are being used, what they are being used for and what problems exist (see the Technology Analysis Form in Chapter 7).

The level of analysis may range from interviewing selected users to a comprehensive survey involving specially designed questionnaires and automatic data collection (automatic recording of telephone traffic, for example).

The output from this survey should indicate the current and projected levels of use of each system and the problems that are currently being experienced in each organizational subgroup.

Identification of high-benefit areas   Based on the results of the previous analysis the OIS team should be able to identify those areas of the organization which could derive immediate benefit from OIS technology. They should specify what problems are currently being experienced in these areas, differentiating between areas that have problems that can be solved without OIS and those that require it (Tasks and Links Report in Chapter 7).

This identification is important. Many trials fail because insufficient attention has been paid to the initial application. Since the introduction of OIS systems includes some disruption to normal routines, it is necessary that the benefits for initial applications be obvious to the user group. The benefits also must be sufficiently high to offset management's concern regarding the associated investment of time and money.

OIS opportunities report   The results of the preceding analyses should be summarized in an OIS opportunities report to senior management. This report should discuss the functioning of the organization and indicate the following:

(1) a description of areas most likely to benefit from OIS;
(2) a list of particular problems OIS should solve;
(3) a description of how OIS will contribute to goal attainment;
(4) proposals for reorganization resulting from the preceding analyses, indicating how OIS fits with organizational functioning;
(5) projections of overall system demands over the next three to five years;
(6) estimates of costs for several alternative OIS systems.

The report should contain sufficient detail so that senior management can decide whether it is worthwhile to continue with the OIS process or if other alternatives are likely to be more cost-effective.

Management review   The process of needs assessment should end with a management review of the opportunities report and a decision regarding the continuation of the OIS program. At this point senior and middle managers have the opportunity to consider the problem of automation of the office in the context of the current organizational configuration and goals. The alternatives raised by the report may also engender further investigation. This is the point where the data from the HIT analysis should be used to generate a new strategic vision of the organization.

Technology Assessment

In this process, detailed technical specifications are not addressed. Rather, the aim is to gain an understanding of the inherent capabilities of the emerging technologies and to assess trends so that systems can be obtained which will fulfil both short- and long-term needs. This is a technical step in the sense that some technical knowledge is required to conduct a realistic assessment of the direction in which office systems and information systems in general are moving.

A first step in the technology assessment program is to become familiar with current vendor offerings and expected future trends through various trade publications, trade shows and conferences (Lientz and Chen, 1981, describe a similar approach).

The next step is to schedule equipment demonstrations with vendors. Direct experience with systems is probably the best method of gaining an understanding of the capabilities of different products. Before venturing into the high-pressure world of the vendors there are system design issues that should be considered. These are expandability, upgradeability, reliability, compatibility, integration and ergonomics (this analysis supplements our discussion of the technical aspects of integration of business systems in chapter 3).

(1) *Expandability*: this describes the system's capabilities to grow with increasing demands. This can involve adding more users or increasing the system's storage capacity. Some systems are inherently limited. For example, if one buys a microcomputer with 12 serial ports it may not be possible to expand the number to 16 to meet increased demands.

(2) *Upgradeability*: this refers to the extent to which a system can be upgraded as new software or hardware become available. Unfortunately, some systems are not standardized. In these cases one must sometimes replace an existing system with a completely new one in order to take advantage of new features that are now possible. This can involve not only considerable hardware and software costs, but extensive user retraining as well.

(3) *Reliability*: in one sense, reliability refers to the degree to which the system will continue to perform its functions without breakdowns. Another aspect of reliability is the time it takes to get the system running after a failure. In some cases this may be more important than the number of breakdowns. Software reliability is concerned with the functions the system performs. In some cases the software itself may contain unexpected errors. For this reason it is strongly recommended that well-proven software be obtained where possible.

(4) *Compatibility*: this is a capacity of the system to relate to other systems. In particular one should investigate how well the system being considered will fit with existing systems and with other systems planned for the future. As yet there are minimal standards for compatibility between computer systems; though this is changing. For instance, most vendors of local area networks assert that their systems are compatible with many hardware products or operating systems. What this means is that a variety of systems can be connected and can communicate with similar products.

(5) *Ergonomics*: factors such as health, safety and ease of use should be considered here. An additional consideration is the functionality of the system in terms of training. The flexibility of the system design, and the extent to which it can tolerate variations in work styles, should also be examined.

A more general problem is determining which technology is appropriate to use. An example of competing technologies is computer-based voice-mail and computer-based text-mail. Both systems offer similar services. The real question is which is appropriate in light of task requirements and environmental characteristics.

Project Schedule

The project schedule describes the proposed projects and presents a timetable, as well as associated budgets and relative priorities.

The level of detail should be sufficient so that each manager affected can include these activities as part of the master plan for his or her unit. Standard project management and cost control techniques apply here. For a good overview of project management techniques we recommend the following books: Love (1980) and Meredith and Mantel (1989).

The Short-range Plan

This one- to two-year plan is aimed at the operational level and focuses on the accomplishment of specific projects. All of the normal project management technologies are applicable here. A unique feature of the OIS implementation process is the pilot trial approach whereby appropriate technologies are installed on a limited basis. This is particularly true during the first three to five years when the organization is adjusting to changes brought about by the new technology. The following remarks are aimed specifically at pilot projects but pertain to ongoing implementations as well. The issues are identified in terms of user-related issues and training.

User-related Issues

Careful planning and consideration of human factors are keys to an office automation implementation strategy and contribute significantly to user acceptance of the recommended system. The following issues should be addressed:

Pre-installation orientation of user group   Group sessions with all potential users of the new system should be conducted by members of the OIS team prior to the installation. The purpose is to educate the user group regarding the capabilities and functions of the technology, and to reduce uncertainty regarding office automation. The sessions should include the direct users and others who will be affected by the proposed changes. If possible, prior to installation, a hands-on demonstration should be arranged on the premises. The session leader should be prepared at this point to answer questions of a technical nature and address workers' concerns and fears regarding the impacts of the new technology (e.g. possibility of job reductions, physical side-effects).

Preparation of revisions to work procedures   Changes to existing procedures will have to be implemented with the introduction of new technology. Both the potential benefits and problems (e.g. security, access to information) that might arise from new automated procedures should be discussed with the user group. Wherever possible the OIS team should begin work with members of the user group to prepare revised procedures for changes in information handling or work flow (e.g. maintenance of electronic files). Otherwise, these changes could result in an increase in the workload so that some staff are overwhelmed with the work or are faced with radically different tasks.

Equipment layout plan   Plans must be made for the details of the physical layout of the new equipment. Work areas may have to be rearranged in order to meet constraints regarding wiring distances, placement of terminals and confidentiality of material displayed on screens. For some equipment, provision for cables, air conditioning and noise reduction will be required.

Assignment of job responsibilities   With the acquisition of new technology it may be necessary to reassign duties or delegate new tasks to current staff. To avoid confusion and unnecessary tension among staff when the new equipment is installed, these issues should be addressed well in advance of equipment delivery. Some of these will be imbedded in the opportunities report.

Training Issues

The objective of the training programme is to provide users with the understanding and skills required to operate the technology effectively. Although training programs will vary with the type of technology, certain key issues must be addressed in any installation. These are discussed in the following sections.

Need for developing expertise within the organization   There are a number of reasons for developing in-house training programs. Training and support for the technology will be required on a continuing basis (e.g. to instruct new users as staff turnover occurs or to provide follow-up to the initial training). The only cost-effective approach is for the organization to be self-reliant in this area, since external training courses (i.e. from vendors or consultants) are expensive and may not meet individual needs. In many cases reference manuals are the only training tool provided by the vendor.

Initial training session   Upon installation of the new technology, training sessions should be scheduled with each member of the user group. These should be conducted on a one-to-one basis, allowing the user to ask questions and to have direct hands-on experience with the new equipment. The length of time allocated for individual training will depend on the type of technology implemented and the characteristics of the individual being trained.

Follow-up assistance training   It is extremely important that the in-house expert(s) be readily available to assist users during the months following installation. In addition to the provision of daily informal support, formal follow-up training sessions should be scheduled with individuals to ensure proper use of the technology.

These follow-up sessions can also help to prevent user frustrations from developing into more serious problems. The idea is to organize group sessions after all participants have had adequate time to use the technology on a regular basis, so that each will feel prepared to contribute his or her ideas and comments.

Reference manual/documentation The OIS team should be prepared to develop their own user manuals, as there may be problems with vendor manuals. It is important that reference manuals be kept simple, compact, and include examples of applications and a clearly organized index and table of contents. For quick and handy prompting of the user's memory, a simple and concise reference card should also be made available. Online assistance should be provided, where appropriate, to guide the user through the operation of the system when problems are encountered (e.g. electronic mail).

Ongoing Assessment

Though our framework does not explicitly identify ongoing assessment as a major step, it is a necessary activity. Ideally the OIS team, management and users will conduct periodic review of various applications to determine if:

(1) the system is functioning according to specification;
(2) the specifications accurately reflect the needs;
(3) the needs have changed.

CONCLUDING REMARKS

A step-by-step planning procedure, which ties together strategic, medium-term and short-term planning, was presented. The role of users and management at each stage of the planning process is an important part of this planning process. One must note that the process is merely a tool. A primary key to success is the vision, commitment and energy of senior management. Without these, any planning process is merely a sterile exercise devoid of meaning. Needs analysis is one other key to a successful planning initiative. In Chapter 7 a methodology for managerial needs analysis is presented.

ACKNOWLEDGEMENT

We acknowledge the support of Microtel Limited who funded the development of a monograph on office automation planning which forms the basis for this chapter.

BIBLIOGRAPHY

Ackoff, R. L. (1959) 'Systems, organizations and interdisciplinary research', *General Systems Yearbook*, **5**, 1–8.
Ackoff, R. L. (1967) 'Management misinformation systems', *Management Science*, **14**(4), B147–B156.
Ahituv, N. and Neumann, S. (1982) *Principles of Information Systems for Management*, Wm. Brown, Dubuque, IO, pp. 167–170.

Akoka, J. (1981) 'A framework for decision support systems evaluation', *Information and Management*, **4**, 133–141.

Argyris, C. (1971) 'Management information systems: the challenge to rationality and emotionality', *Management Science*, **17**(6), B275–292.

Asner, M. and King, A. (1981) 'Prototyping: a low-risk approach to developing complex systems', *Business Quarterly*, Autumn, pp. 30–34.

Asner, M., King, A. and Drake, R. G. (1981) 'Prototyping: a low-risk approach to developing complex systems (Part 2: Methodology)', *Business Quarterly*, Winter, pp. 34–38.

Barnard, C. I. (1938) *Functions of the Executive*, Harvard University Press, Cambridge, MA.

Barrett, D. and Oman, K. (1983) 'Micro-based performance and productivity for managers and professionals', *Business Quarterly*, Spring, pp. 61–69.

Bowman, B., Davis, G. and Wetherbe, J. (1983) 'Three stage model of MIS planning', *Information and Management*, **6**, 11–25.

Bryce, M. (1983) 'Information resource mismanagement', *Infosystems*, No. 2, pp. 89–92.

Buss, M. D. J. (1983) 'How to rank computer projects', *Harvard Business Review*, January–February, pp. 88–125.

Canas, A. J. (1982) *Bibliography on Office Information Systems*, CECIT Working Paper.

Carter, J. C. and Silverman, F. N. (1980) 'Establishing an MIS', *Journal of Systems Management*, January, pp. 15–21.

CECIT (1982) *Development of a Set of Methods, Guidelines and procedures to Assist in the Specification and Evaluation of Office Communication Networks and Services, as they Pertain to Office Automation*, University of Waterloo, Waterloo, Ontario, September.

CECIT (1982) *An Annotated Review of Literature on the Specification and Evaluation of Office Communication–Information Systems*, University of Waterloo, Waterloo, Ontario, June.

Easterby-Smith, M. and Davies, J. (1983) 'Developing strategic thinking', *Long-Range Planning*, **16**, 39–48.

Elton, M. C. J. and Carey, J. (1981) 'Implementing interactive telecommunications services', *Alternate Media Centre Report*, New York University, New York.

Ghosh, B. C. and Nee, A. Y. C. (1983) 'Strategic planning—a contingency approach—Part 1: The strategic analysis', *Long-Range Planning*, **16**, 93–103.

Irving, R. H. and Munro, S. (1983) 'Facilitating the adoption of office technology', Proceedings of ASAC Conference.

Kay, R. H., Szyperski, N., Horing, K. and Bartz, G. (1980) 'Strategic planning of information systems at the corporate level', *Information and Management*, **3**, 175–186.

Keen, P. G. W. (1988) *Competing in Time*, Ballinger, Cambridge, MA.

Kelley, C. D. A. (1983) 'The three planning questions: a fable', *Business Horizons*, March–April, pp. 46–48.

Kelley, H. H. and Thibaut, J. W. (1969) 'Group problem solving', in G. Lindzey and E. Aronson (eds.), *The Handbook of Social Psychology*, **4**, 2nd edn, Addison-Wesley, Toronto, pp. 1–101.

King, W. R. (1983) 'Planning for strategic decision support systems', *Long-Range Planning*, **16**(5), 73–78.

Lahr, M. L. (1983) 'Interactive planning—the way to develop commitment', *Long-Range Planning*, **16**(4), 31–38.

Leonard-Barton, D. and Kraus, W. A. (1985) 'Implementing new technology', *Harvard Business Review*, November–December, pp. 102–110.

Lientz, B. P. and Chen, M. (1981) 'Assessing the impact of new technology in information systems', *Long-Range Planning*, **14**(6), 44–50.

Lorsch, J. W. and Lawrence, P. R. (1965) 'Organization for product innovation', *Harvard Business Review*, January–February, pp. 109–122.

Love, S. F. (1980) *Planning and Creating Successful Engineering Designs*, Van Nostrand Reinhold, New York.

Maier, N. R. F. and Maier, R. A. (1957) 'An experimental test of the effects of "developmental" vs. "free" discussions on the quality of group decisions', *Journal of Applied Psychology*, No. 41, pp. 320–323.

Maier, N. R. F. and Solem, A. R. (1952) 'The contribution of a discussion leader to the quality of group thinking', *Human Relations*, No. 5, pp. 227–288.

McConkey, D. D. (1981) 'Strategic planning in nonprofit organizations', *Business Quarterly*, Summer, pp. 24–33.

McFarlan, F. W., McKenney, J. L. and Pyburn, P. (1983) 'The information archipelago—plotting a course', *Harvard Business Review*, January–February, pp. 145–156.

McKeen, J. D. (1983) 'Successful development strategies for business application systems', *MIS Quarterly*, **7**(3), 47–65.

McLean, E. R. and Soden, J. V. (eds) (1977) *Strategic Planning For MIS*, Wiley-Interscience, New York.

Meredith, J. R. and Mantel, S. J., Jr (1989) *Project Management: a Managerial Approach*, John Wiley, New York.

Mertes, L. H. (1981) 'Doing your office over—electronically', *Harvard Business Review*, March–April, pp. 127–135.

Mittra, S. S. (1983) 'Information system analysis and design', *Journal of Systems Management*, April, pp. 30–34.

Morton, M. R. (1983) 'Technology and strategy: creating a successful partnership', *Business Horizons*, January–February, pp. 44–48.

Mumford, E. (1981) 'Participative systems design: structure and method', *Systems, Objectives and Solutions*, **1**(1), 5–19.

Oliver, A. R. and Garber, J. R. (1983) 'Implementing strategic planning: ten sure-fire ways to do it wrong', *Business Horizons*, March–April, pp. 49–51.

Paul, R. N., Donavan, N. B. and Taylor, J. W. (1978) 'The reality gap in strategic planning', *Harvard Business Review*, May–June, pp. 124–130.

Porter, M. (1985) *Competitive Advantage: Creating and Sustaining Superior Performance*, Free Press, New York.

Tapscott, D. (1982) *Office Automation: A User Driven Method*, Plenum Press, New York.

Tapscott, D., Henderson, D. and Greenberg, M. (1985) *Planning For Integrated Office Systems, A Strategic Approach*, Holt, Rinehart & Winston of Canada, Toronto, Canada.

Trigon Systems Group Inc. and CECIT (1982) *Report to Department of Communications*.

Uhlig. R. P., Farber, D. J. and Bair, J. H. (1979) *The Office of the Future*, Elsevier North-Holland, New York.

Woodward, J. (1958) *Management and Technology*, Her Majesty's Stationery Office, London.

Chapter 7
NEEDS ANALYSIS FOR OIS: A MANAGEMENT PERSPECTIVE

Once an appropriate planning process is in place, the requirements (or needs) analysis phase is the most crucial step in the process of introduction of OIS to an organization. Unfortunately it is also the least scientific step. Not only are managers forced to balance conflicting demands of various departments, there are also a variety of competing methodologies each with their biased proponents. Compounding these difficulties is the mass of information that typical needs analysis procedures generate. Furthermore, since the organizational changes can be profound, and since the resulting benefits may be hard to realize, it is tempting to delegate the needs analysis process to a third party. This provides a convenient scapegoat in the event that things go wrong.

WHY MANAGERS SHOULD BE INVOLVED IN REQUIREMENTS ANALYSIS

Though managers often delegate much of the responsibility for needs assessment to external consultants or in-house 'professionals', in our view this is unwise.

A major reason for strong management involvement in this phase is that an OIS does much more than provide support for organizational tasks (see Chapters 3 and 4). These systems integrate organizational processes, change the degree of coupling between tasks, workgroups and departments and alter the very nature of work. In so doing, they (OIS) change how employees perceive their roles in the corporation and the view the corporation has of its employees. Though the degree to which these effects are manifested will vary from one organization to another, they are present to some degree in each implementation. Consequently, the requirements analysis phase of OIS is the first step in a process of organizational redesign. We argue that managers must take a leading role in this process.

While a key reason for managerial involvement is the profound nature of the potential changes, there are other equally good reasons. Middle levels of management contain individuals who are most likely to understand the business issues in their organization at an operational level and who have the perspective of the organization as a whole. It is this high level of experience, knowledge and insight that is required when redesigning the organization through the use of OIS.

Most external consultants (see Case 1) lack exactly these qualities and therefore spend considerable time (and organizational expense) acquiring sufficient information to make reasonable judgements. Though in-house professionals have good technical training in systems analysis and design, and are more familiar with the organization than external consultants, many lack sound business knowledge, and are at too operational a level to have an appreciation for general business concerns (see Case 2). In particular, many of the design decisions have strategic and operational implications which require management input. In our experience, in-house people frequently do not seek this input. This is particularly true when the information systems department drives the OIS. When functional levels of management are leading the OIS project these problems are minimized.

CASE 7.1: Confusion by generalizations

A financial institution hired a major consulting firm to conduct a needs analysis. After spending $50 000 the consultants developed a series of general strategies and designed an overall system architecture. While the overall ideas were good, management complained that the final result was too general to be of any use, and that the information was presented in a way that was difficult to understand. This type of problem is typical of those encountered when prominent consulting firms are retained to do an analysis of requirements for OIS.

We call this the 'confusion by generalizations' problem (CBG). Managers face several difficulties when this type of problem arises. First, the information provided may be appropriate for senior management making broad strategic decisions, but is inappropriate for middle management who must decide on specific applications. Second, many of the consulting firms use a proprietary methodology which in effect transfers information about the firm to the consultant, who then uses that as a basis for deciding the organization's requirements. To the extent that the information transfer has taken place accurately, the resulting system may be appropriate. However, since most organizational personnel will not understand the process whereby the recommendations were developed, they are likely to mistrust the output from that process. In some cases they are wise to do so. The president of a small manufacturing firm related an incident where the firm hired a well-known company to develop an information system. After spending $30 000 for the study and another $50 000–70 000 for the system they found that the final product was of little use. Although the company did not go bankrupt, in the words of the president, 'it was nip and tuck for about 12 months.'

Presumably we have established why managers must be involved in requirements analysis for OIS. We now turn to the nature of this involvement. To provide a context for our remarks we first review current approaches to needs/requirements analysis and then present the role of management in this process, and a method to facilitate their effective involvement.

CASE 7.2:   Confusion by detail

A director of a staff group in a large university wanted to automate his operations. These functions consisted of processing grant applications, supplying information to faculty members and tracking the use of grant funds. He employed an analyst from the computer services department to examine departmental work patterns and to produce a report which could be used for system planning.

After six months of data collection and time-consuming analysis the director received a detailed task breakdown which described in minute detail all of the activities which occurred in his department. When we talked to him the report was gathering dust. In the director's words, 'She [the analyst] gave us an incredibly detailed description of what we were doing 6 months earlier. What I wanted to know was what should we be doing and how could computer technology facilitate these activities!' The level of detail was too great for the director's decision making needs and did not address his real concerns.

We characterize this common problem as 'confusion by detail' (CBD). In this case the manager, who is looking for some general guidelines for the areas most amenable to automation, is deluged with huge volumes of minute detail that obscure the overall patterns. This type of problem is most apt to occur when one has traditional data processing personnel (systems analysts) conduct a needs analysis. This is typical of someone from within the organization who is too oriented to their operational level of concern and who cannot put general business needs into perspective.

APPROACHES TO NEEDS/REQUIREMENTS ANALYSIS

In the following sections we provide a descriptive tour through the various approaches to needs analysis. This tour is divided into two stages: (1) underlying models and (2) domains of application. The first stage addresses the underlying models of organizational structure from which various methodologies are derived; the second discusses methodologies under the subheadings MIS methodologies and OIS methodologies.

Underlying Models of Organizational Structure

During the past 10 years several approaches to defining the requirements for office information systems have been developed. Hirschheim (1985) categorizes them as *analytic* or *interpretivist*. The analytic perspective views offices as highly structured, task-oriented and rule-dominated. The interpretivist perspective, on the other hand, views the office as a social setting where human relations, power, myth and symbols are dominant characteristics.

The analytic approach has been discussed by a number of authors. The major criticisms of it are that it focuses on the superficial structure of tasks and activities and ignores the deep structure of political action and human emotions. This criticism seems valid in many cases. Markus (1983) demonstrated the dangers inherent in implementing MIS while ignoring the power and politics of the organization. However, in many settings a simple analytic perspective can produce adequate results. If the organization is in a period of political tranquillity, and if the major activities of organizational members are highly structured (e.g. an insurance company or small manufacturing firm) an analytic perspective can be appropriate.

The interpretivist perspective, by focusing on the human aspects of an organization, provides insight into the cultural and sociopolitical realms. These insights can be valuable; however there is a problem: most of the sociotechnical approaches identify social problems but provide little help in incorporating these insights into the design of an information system.

Hirschheim quite properly notes that a mixed methodology incorporating elements of both approaches may be appropriate. From our point of view the analytic perspective provides the data and structure for a needs analysis (e.g. the content) and the interpretivist perspective provides the context in which these data are collected (e.g. the process). Data collected using a sociotechnical process reflect the human side of the organization while capturing the structure.

As we will show in the section on the manager's role in the requirements analysis process, there is a third perspective, which we label the integrationist perspective. This approach uses an analytic organizational model and an interpretivist (sociotechnical) process of data collection to gather requirements data which reflect the collective wisdom of the organizational participants while at the same time providing sufficient structure for targeting in-depth needs analysis.

In addition to the general perspective, one must also consider how the methodology is used, or its domain of application.

Domains of Application

There are two domains of application relevant to OIS. The first consists of MIS requirements analysis methodologies that have been developed to identify requirements for transaction processing systems and more recently for management information systems. These methodologies are well tested by extensive field use, and their strengths and weaknesses are well understood. The other methodologies have been used for office automation and are more recent. These methods attempt to capture the interrelationships between office activities and tasks. Some of the methodologies stress organizational culture and interpersonal relationships from a sociotechnical perspective. These methodologies are less well tested than the MIS methodologies discussed earlier.

MIS Methodologies

Bowman, Davis and Wetherbe (1983) present a good overview of MIS requirements analysis methodologies. They also discuss problems associated with MIS planning and emphasize that the MIS plan must fit the needs and priorities of the organization. If the plan is based only on needs and priorities of users this macro perspective may be lost. They note that there are a large number of competing methodologies which are similar but not equivalent, and that each method has its advantages and disadvantages. The categories of planning and needs assessment methodologies described by Bowman *et al.* are summarized in Table 7.1.

The eight methodologies represent a broad spectrum of MIS applications. The first (Strategy Set Transformation) is aimed at the strategic level of the

TABLE 7.1   Summary of MIS requirements analysis methodologies

(1) *Strategy set transformation:*
  define a strategy set for organization;
  transform this to a set for MIS;
  focus only on strategic planning;
  needs articulated organization goals.
(2) *Business systems planning (BSP):*
  developed by IBM;
  conducted by BSP planning team of users and MIS personnel;
  phase 1: focuses on understanding the organization;
  phase 2: develop long-range plan for design, development and implementation;
  a top down approach.
(3) *Critical success factors (CSFs):*
  developed by Zani and Rockart;
  four primary sources of CSFs are:
  (a) industry factors—competitive strategy and industry position,
  (b) geographical location,
  (c) environmental location,
  (d) temporal factors;
  uses a series of interviews, but less than BSP.
(4) *Business information analysis and integration technique:*
  developed by Burnstine;
  uses seven questions to develop a normative set of information requirements;
  focus is on orders and customers;
  still experimental.
(5) *Ends–means analysis:*
  developed by Wetherbe and Davis;
  focus is on ends (outputs) and means (inputs).
(6) *Return on investment (ROI):*
  cost–benefit analysis technique.
(7) *Chargeout:*
  fee schedule for MIS services (traditional MIS).
(8) *Zero-based budgeting:*
  developed by Peter Phyrr;
  conceptually reduce all MIS activities to zero base;
  no development or maintenance of MIS; applications are analysed on benefits and resource support requirements; useful in identifying outmoded applications.

organization. Business System Planning (No.2) attempts to be a comprehensive methodology. Others such as ROI focus on various aspects of the complete system justification, analysis, design and implementation process. Readers who want more information should read the Bowman, Davis and Wetherbe article.

OIS Methodologies

Starting in the early 1980s several OIS analysis methodologies were developed. We do not intend to provide an extensive review here. Those readers who desire more information should read chapter 4 in Hirschheim's (1985) book on OIS.

Hirschheim notes that there is a common theme in many of the newer methodologies which explicitly recognize the need for extensive user involvement in the requirements specification and system design process. Earlier information system methodologies, such as Enid Mumford's ETHICS, explicitly incorporate sociotechnical systems thinking in the process. However, the OIS/OA methodologies developed in the late 1970s and early 1980s were notable for their insistence on user involvement. Among the methodologies are Tapscott's (1982) user-driven design methodology, Checkland's (1981) soft systems methodology, Pava's (1983) sociotechnical design methodology and MOBILE (developed by Dumas *et al.*, 1982). All of these approaches provide the possibility of tailoring OIS to the needs of individual organizations.

Though we recognize the strengths of these methods, in our experience there has been little effort expended on generating a methodology which will provide condensed information to management early in the assessment process, so that management can control the process of systems development. In the next sections we show a process which addresses this issue.

MANAGERS' ROLE IN THE NEEDS ANALYSIS PROCESS

Our concern for the problems managers have in dealing with the needs assessment process came as a result of our own research, and as a consequence of working with managers and consultants over the past 10 years. During this period we have heard managers complain about the difficulty of determining the proper allocation of investments in OIS. Another issue concerns the choice of personnel who should be involved in the initial planning process. In addition, the assistance that managers need in order to make decisions is often not available, or is inappropriate to their needs. These might seem like trivial issues to some, but to practising managers they are serious concerns. Among the questions most frequently asked are the following:

Where should I concentrate my resources?
Which groups and individuals should be involved in the planning and design process?

How can I relate the use of the technology to the main work processes for which I am responsible?

Typically, middle-level managers in large corporations have little input to overall corporate (or macro) strategy. At the same time they have some control over their own departmental (or micro) strategies. Furthermore, most managers want data summarized in a form that is quickly understandable to them. They will not wade through a detailed analysis to determine if it makes more sense to spend money on one or another department. They will, on the other hand, be ready to make decisions if the impact of technology on the various departments can be presented in a comprehensible and abbreviated form, and if the process of analysis is understandable and looks reasonable.

Before addressing new technology there are several detailed questions that require answers. These are:

(1) Are the resources of the subordinate workgroups appropriately divided?
(2) Are the links between workgroups working smoothly?
(3) Are there better ways to organize these workgroups to obtain greater effectiveness and efficiency?
(4) Are we focusing on useful organizational tasks?
(5) How can we ensure that important inter-workgroup links are maintained?
(6) What are the key problems that are currently being experienced with existing systems?
(7) How can we accommodate possible future needs?

Once these issues have been addressed one can attempt to answer other more technology-oriented questions such as:

(1) Are there technologies which can be used to facilitate the key tasks of this set of workgroups taken as a whole?
(2) How can we set priorities for system development?
(3) In redesigning a particular system, which workgroups should be consulted?

When management has delegated control of the needs analysis process to external 'experts' these questions are generally either not asked, or are not answered properly. In the following section we show how appropriate answers to these questions can be generated, and how management can use this information to control the needs analysis and system design and implementation process. We begin with a simple model of the office and discuss the data that must be collected to answer the key questions. We then describe a process whereby these data can be collected.

A General Model

The unit of analysis for our model (see Figure 7.1) is the workgroup. This consists of a supervisor (or manager) and his or her immediate subordinates. Each workgroup has a set of key tasks, and links to other workgroups to accomplish those tasks.

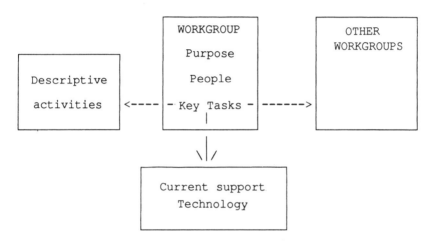

FIGURE 7.1   The general model

The key tasks are presumably related to the purpose/mandate of the workgroup and represent those major tasks which define its overall responsibilities. Thus, if these key tasks are done *well*, the group will be seen as performing *well*. Individuals belonging to the workgroup can be associated with the key tasks of the workgroup in terms of their primary and secondary responsibilities.

For a particular key task, there will likely be links to other workgroups. These links represent channels of communication and information flow that are necessary to the performance of the task. At our level of analysis it does not matter if the link is a flow of paper or face-to-face communication. The important issue is whether or not the links exist, and the problems with these links. The initial assumption is that a new OIS must at least maintain these existing links.

In addition to the links we can identify support technology which is currently being used to facilitate task accomplishment. This technology may be computer-based, though it need not be. It can range from telephones and central copying facilities to custom databases and video conferencing. It is important to identify both the technology which supports each task and the quality of that support. This information is central to answering questions relating to possible opportunities for technical support.

Each task may contain a number of activities or subtasks. We try to include several descriptive activities to uniquely identify each key task.

When the preceding information has been collected for each workgroup in a department, division or corporation, one has an overall picture of the key tasks, levels of technological support and degree of linkage between workgroups. This information provides the basis for effective management control of the needs analysis process and their ongoing control of the design and implementation of OIS. The following example illustrates the concepts discussed in this section.

A Simplified Example

Best Electronics Inc. (BEI) makes computerized toy clowns that walk, talk and do card tricks. BEI has a production department consisting of a manager and 12 subordinates, a marketing department consisting of a marketing manager and 15 salesmen, a design department with five junior designers and one senior designer, and an executive team consisting of the president, controller, and accounting manager. The accounting manager is supported by five accounting clerks. The BEI structure appears in Figure 7.2.

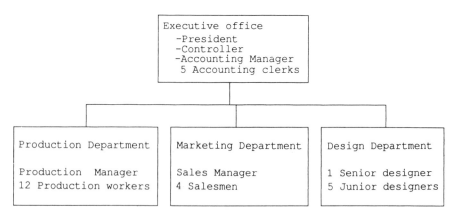

FIGURE 7.2   Organizational structure of Best Electronics Inc.

The president of BEI (Joe Smith) has decided to install an OIS to integrate his three departments and executive office, and to streamline the handling of communication and information. Currently the organization's only support technology consists of a telephone system, electronic typewriters and a simple PC-based CAD system for the design office.

Joe's major uncertainties related to integrating a new system with existing organizational functions and targeting those areas most in need of support. Since his organization was small he did not conduct an organizational scan as described in Chapter 2, though he did commission a needs analysis. Before describing the process (the HIT technique) let us look at the information Joe obtained, and discuss how he was able to use it to help him manage the requirements analysis and implementation process. The details of the key tasks, use of technology and links for each workgroup are contained in the workgroup reports (see Supplement 7.1). These reports contain information on the workgroup, its mandate and key tasks, as well as outline details of the workgroup members and their task responsibilities. In addition, the second page of each workgroup report contains information on the key links to other workgroups for each key task, and use of support technology for each key task.

The BEI needs analysis can be discussed in terms of four reports: the Linkage Report (Table 7.2); the Linkage Summary Report (Table 7.3); the System Use Report (Table 7.4) and the Links and Support Report (Table 7.5). Each of these reports refers to key tasks by number; the task descriptor can be identified by referring to the workgroup Reports in Supplement 7.1. Once the implications of the BEI data have been discussed, we will show how these data were collected.

The first report Joe perused was the Linkage Report (Table 7.2). He immediately noticed some organizational discrepancies. For instance the

TABLE 7.2   Linkage report by workgroup and task

Workgroup	High links to	Medium links to	Low links to
Executive office	Design department (3,5)	Production department (4,5) Marketing department (4,5)	Production department (1,3) Marketing department (1,3) Design department (1)
Production department	Executive office (4)	Design department (5)	Marketing department (1)
Marketing department	External clients (1) External other (2,3)	Executive office (1,3) Production department (1)	
Design department	Marketing department (2)	Production department (1,2) Marketing department (1)	Executive office (3)

TABLE 7.3   Linkage summary report

Links to:	Internal links to:			
	Executive office	Production department	Marketing	Design
Executive office	1,2,3,4,5	1,3,4,5	1,3,4,5	1,2,3,4,5
Production department	4	1,2,3,4,5,6	1	5
Marketing	1,3	1	1,2,3,4	1
Design	3	1,2	1,2	1,2,3

TABLE 7.4   System use report by system, task and workgroup

System	Executive office	Production department	Marketing	Design
Telephone	1,4,5	1,2,4	1,2,3,4	1,2,3
Typewriter	1,2,3,4,5	1,2,4	1,4	3
Copier	1,2,3,4,5	1,4	1,4	
PC–CAD			1,2,3	

TABLE 7.5   Links and support report by task

Links	Support	
	High	Low
High	DD-2	EO-3,5
		PD-4
		MD-1,2,3
	DD-1	EO-1,3
Low		PD-1
		DD-3

DD-2 = Design Department, task No. 2

executive office indicated that they had a high link to the design office on their key tasks 3 and 5. The design office did not indicate any high links with the executive office. Joe marked this relationship for further investigation. In fact Joe quickly discovered that there were no reciprocal high links. In other words there were no two workgroups where both indicated they had high links with each other on their respective key tasks. In fact for the marketing department, the high links were external to the organization. In this latter case Joe decided that he should check with marketing. For the other three workgroups, Joe decided that there was likely insufficient internal coordination.

So far the data had stimulated Joe to ask questions regarding organizational design. Should the organizational structure be changed before introducing OIS? Joe planned to use these data to promote a dialogue among the three department managers and their staff. The linkage Summary Report (Table 7.3) shows all links to other workgroups (high, medium and low). Joe noted that each workgroup had at least one task (e.g. task 5 for the executive office) that had no external links. Could these tasks be totally independent? If so, could they be supported in isolation from other systems? Would this be a wise course of action?

Before arranging a meeting with his staff, Joe turned to the Systems Use Report (Table 7.4). This report showed that three of the four systems were used by all four departments. The PC-CAD system was used only by the design department.

Joe also noted that the PC system was being used only for CAD applications. The other obvious fact was that there was little use of office technology. While this could involve considerable investment, it also meant that there were few existing systems to replace. The Links and Support Report (Table 7.5) showed this situation clearly. Only the design department had high support and only for tasks 1 and 2 (DD-1, DD-2).

The four quadrants of Table 7.5 helped Joe determine general strategies. For example DD-2 had high links and high support. If BEI wanted to change the technological infrastructure supporting this task, it is clear that a careful and detailed systems analysis must be done to ensure that all links were maintained, and that the new system was at least as useful as the old one. In this case the support was a PC-CAD system. The main question was whether the existing files could be easily transferred to the new system. For DD-1 there were low links; therefore the decisions could be kept mainly within the design department.

For those tasks with high links and low levels of support, the question was mainly one of maintaining and improving existing links. During the process of detailed design and implementation of a new system, extensive consultation with other workgroups must take place. Finally, for the tasks in the low–low quadrant, technical support can be treated as a stand-alone proposition. There may be overriding technical factors, but from an organizational perspective as it exists now there is little reason for extensive consultation.

The information presented to Joe should stimulate him to think first about organizational design issues. Should the workgroups be more tightly linked on various tasks; on which tasks or workgroups should we place strategic priority? When Joe meets with the other department heads they can discuss concrete aspects of their activities and how they should be organized. Later on, as a detailed plan is prepared and implementation progresses, Joe can track the progress in terms of key tasks and key organizational links. Furthermore, Joe and his staff can quickly determine how current systems are perceived by turning to the comments in the workgroup reports. For instance, it seems clear that everyone except the design department is dissatisfied with the current telephone system. This may be an area that BEI wishes to address prior to developing other OIS capability.

The data Joe used were assembled using the Higgins–Irving technique (HIT). In the following section we describe how HIT works.

HIT: Applying the Technique

Earlier we discussed the analytic and interpretivist approaches to developing needs analysis methodologies. Most writers, no matter what their perspective, agree that an effective methodology must combine elements of both viewpoints. In developing the Higgins–Irving technique we formulated an analytic organizational model (see Figure 7.1) and combined it with a sociotechnical process for data collection. As a result structured data are collected in the context of the existing sociopolitical environment of the organization. These data provide management and employees with a view of the existing organization and lead them to assess the existing structure prior to implementing computer-based systems.

It is important when using the HIT process to remember that it provides information for the initial stage in planning. HIT helps management make clear decisions on priorities for OIS and target areas for application. At the detailed system specification stage (and subsequently) information derived from HIT is useful in determining who should be involved in the analysis process and in identifying the systems and tasks to be studied in each target area.

We discuss the process of applying HIT in terms of the six steps involved. Where necessary we identify different courses of action depending on whether one is an external consultant or an employee of the organization. The six steps are as follows:

(1) the preliminary assessment;
(2) design of the HIT forms;
(3) the initial interviews;
(4) the verification rounds;
(5) production of summary reports;
(6) presentation to management.

Step 1: The Preliminary Assessment

The preliminary assessment involves meeting with senior corporate personnel; the process of the assessment varies depending on whether one is an external consultant/researcher or an internal manager/analyst. The process for each is discussed below.

External consultant/researcher   Ideally, you would gain entry to the corporation through a senior vice president or staff person. This should generally not be a data processing (DP) or information systems (IS) person, as our experience leads us to believe that these people usually have a vested interest in the *status quo*. Once entry has been gained, however, you need the cooperation of a DP or IS person to act as an information resource. Ideally, you would also have access to a senior staff person. These resource people provide information on the workgroups and the technology, and will usually help frame the data collection instruments in the language of the corporation. In addition, it is helpful to have an internal person schedule the interviews. The most senior executive available should write a letter to inform all of the personnel who will be involved in the study of the reasons for the interviews, and the time requirements necessary. This should ensure a reasonable level of cooperation and dispel any rumours.

Internal manager/analyst   The activities are much the same as for an external consultant in that a list of workgroups, etc, must be prepared, and senior management must be solidly behind the project if it is to succeed. The main difference lies in two areas. First, an internal analyst or manager must assess his or her own credibility within the organization. If there is any problem an external person should be hired to collect these data. Second, there must be sufficient time to do a proper job. This is much more of a problem for internal people since other 'high-priority' activities are likely to intrude.

One main advantage of an internal person is that he or she is in the best position to assess the quality of information received. In addition, he or she is able to assess the possible uses of the HIT data and to judge the most acceptable formats in which to present these data to management. Finally, competent internal individuals may be able to draw more elaborate conclusions from the HIT data than will external consultants.

Outputs from the preliminary assessment   Assuming that a decision is made to proceed with the HIT methodology, the following are the outputs that should result from the preliminary assessment:

(1) A list of workgroups and workgroup members. This list will be developed in conjunction with the internal liaison and in consultation with other staff and line personnel. It is important to be very careful in developing this list, as it is the basis for the workgroup interviews and for the linkages data.
(2) A list of support technology. This list is again developed in conjunction with the internal liaison and with substantial input from data processing, word processing, OA, and related departments. It is important to have

a comprehensive list of existing support technologies as these will be evaluated in terms of how well they support key tasks.

(3) A schedule of workgroup interviews. The internal liaison should be responsible for developing this list. It is important that a separate interview be conducted with the supervisor of the workgroup, followed by an interview with the other workgroup members. Groups of larger than seven or eight people do not seem to work as well, so if the workgroup is large you may wish to split it into several subgroups.

Step 2: Design of the HIT Forms

In order to be able to collect the data three forms—the Functional Analysis Form (FAF), the Task Linkage Form (TLF), and the Technology Analysis Form (TAF)—must be tailored to the organization. Each form is discussed in the sections which follow.

The functional analysis form   The Functional Analysis Form (FAF), shown in Table 7.6, records the basic descriptive information from which the data regarding links and use of technology are derived. Each item is discussed below (for comparison see the workgroup reports in Supplement 7.1).

Item 1 records the workgroup name, the date of the interview, and the names of the interviewees. Item 2 records a brief statement of the group's purpose or mandate. It should define 'What the group is all about.' Item 3 records the key tasks the group must accomplish. For each task descriptive activities which provide insight into the task are also recorded. Item 4 records the names and titles of the workgroup members and their major task responsibilities.

The Task Linkage Form   The Task Linkage Form (TLF), shown in Table 7.7, is used to relate the key tasks of one workgroup to other workgroups with which they interact. The information collected here identifies the closeness of the coupling between workgroups. This information is useful when planning changes to the organizational structure and when introducing OIS. It provides an overview of the potential impact of any changes on intergroup, task-related communication.

The workgroup name and the date of the interview are recorded at the top of each form. For each workgroup listed on the left-hand side of the TLF, the interviewees are asked to indicate if there is a link with any of the tasks recorded on the FAF. For those tasks where links exist, the importance of the link is recorded by writing the number of the key task from the FAF under the appropriate category (low, medium, high). There is also room to record comments on the nature of the link, problems and concerns. For example,

TABLE 7.6 Functional Analysis Form

(1) WORK GROUP: _____   DATE: _____/_____/1991
    Interviewees: _____
    _____   _____
    _____   _____
    _____   _____
                                    _____

(2) MAIN FUNCTIONS OF WORKGROUP: _____
    _____
    _____

(3) KEY TASKS   ACTIVITIES
    1. _____   _____
    2. _____   _____
    3. _____   _____
    4. _____   _____
    5. _____   _____
    6. _____   _____

(4) WORKGROUP MEMBERS   MAJOR TASK RESPONSIBILITIES
    1. _____   _____
    2. _____   _____
    3. _____   _____
    4. _____   _____
    5. _____   _____
    6. _____   _____
    7. _____   _____
    8. _____   _____

TABLE 7.7 Task Link Form

WORKGROUP: _____   DATE: ____/____/1991

Group	Importance of link			Problems and concerns
	Low	Medium	High	
Executive office	_____	_____	_____	_____
Production department	_____	_____	_____	_____
Marketing department	_____	_____	_____	_____
Design department	_____	_____	_____	_____

if interviewing the executive office, one would write 'executive office' on the first line where the workgroup title is requested and fill in the date of the interview. Subsequently, interviewees are asked to indicate how important a link (if any) there was with the production department for each of the key tasks. A link is defined to exist if one group must interact with another workgroup to accomplish a key task properly.

A link of *high* importance is one in which this interaction is absolutely essential to the completion of the task. A *medium* link is one in which the interaction is important but not absolutely essential. A *low* link exists when the interaction is helpful, but not necessary. At the time of recording the links it is useful to record any problems or concerns that the interviewees raise regarding each link. These concerns might range from simple communication problems involving difficulties in contacting the other party, to structural problems such as differing or conflicting roles.

The Technology Analysis Form   The Technology Analysis Form (TAF), shown in Table 7.8, ties the currently available technology to the key tasks of each workgroup. The data from this form provide information on the level of task support available, and ties key tasks to currently available technology. Such information is useful in assessing levels of support and in planning for future systems.

TABLE 7.8   Technology Analysis Form

WORKGROUP: _____   DATE: _____/_____/1991

Technology	Usefulness of technology			
	Low	Medium	High	Problems and concerns
telephone	_____	_____	_____	_____
Typewriter	_____	_____	_____	_____
Copier	_____	_____	_____	_____
PC–CAD	_____	_____	_____	_____

The list of technologies developed during the preliminary assessment is used for the left-hand side of the form. For each technology (or software package) the interviewees are asked to indicate both the key tasks supported, and the usefulness of that support. This is done by writing the task number from the FAF under the appropriate column indicating high, medium or low usefulness. Comments, suggestions and concerns associated with each task/technology combination are also collected.

Step 3: The Initial Interviews

The initial interviews are conducted by workgroup and consist of two parts. The first interview is with the workgroup leader and provides preliminary data on the workgroup, purpose, key tasks and responsibilities of workgroup members. If time permits, data can be collected on task linkages and the use of available technology by task, but this is not absolutely essential. The second interview is normally (ideally) done with the workgroup members in the absence of their superior. This facilitates the free exchange of information. In this second session the information provided by the leader is verified (i.e. the tasks and responsibilities of group members) and the links with other workgroups and the use of technology by task are identified through the use of the three forms previously discussed.

Step 4: The Verification Rounds

The verification rounds take place subsequent to the completion of the initial interviews. The point of this exercise is to verify that the links and tasks are seen by different workgroups in the same way. To this end the information on links is compiled to see if any natural groupings fall out. These groupings can then be used as a basis for the scheduling of the groups for the verification rounds. Alternately, one can check with the organizational liaisons and jointly determine the composition of the group that will comprise each verification session. In practical terms, having three to five workgroups present for any session seems to be the best; a smaller number is an inefficient use of everyone's time and a larger number makes the meetings unwieldy. Sometimes one or two groups will have so many links that they should be represented at several interviews. To accomplish this effectively it is useful to have a representative of each group at each verification round. During each verification round it is important to have the key tasks for each workgroup displayed prominently. A flip chart with group names and key tasks is very helpful. During the session the investigator proceeds by workgroup; first describing the key tasks and ensuring that everyone understands them. The links to other workgroups are then discussed. Any changes made in either the tasks or in the links are recorded, and the original data forms and the database are changed to reflect this new information. At the completion of the verification rounds a database will have been developed that reflects a consensual view of the key tasks, links and use of technology for every workgroup. In addition, the participants will have a common understanding of this database.

Step 5: Producing the Summary Reports

HIT generates a simple yet rich database. The following reports are useful:

(1) a Workgroup Report;
(2) a Linkage Report;
(3) a Linkage Summary Report;
(4) a Linkage Comments Report;
(5) a System Usage Report;

(6) a System Comments Report;
(7) a System Summary Use Report;
(8) a Links and Support Report.

These reports provide the basis for discussing the data with management. Examples of selected reports are presented in Supplement 7.1.

Though all the preceding reports can be produced either by manual analysis or by use of commercial statistical analysis packages it is preferable to have a tailored package to run HIT data. Not only does this simplify the production of the reports listed above; it also simplifies the data collection and verification. For example, with a HIT database package one could take a portable computer to the initial interviews, enter the workgroup data, and give each participant a copy of the data to review at the end of the session. This allows an extra verification of the data to take place. Similarly, taking a portable computer to the verification rounds allows quick production of the revised reports.

We feel that this speed is essential. By providing quick feedback the whole process is accelerated and participants' help is solicited while they are still 'involved' in the process. When weeks go by before feedback takes place, people 'cool off' and tend to lose interest in the process.

Step 6: Presentation to Management

The group to whom the data are presented should be a mixed management team consisting of both line and staff. Ideally, the manager of OIS or DP will also be present. The individual (e.g. analyst, manager or researcher) presenting the information should be prepared to chair the meeting and ask some provocative questions to stimulate the group.

When the data contained in the reports previously discussed are presented, it should be made clear to management that these data are for their use in stimulating them to ask more sophisticated questions, or to help them target areas for OIS or to set priorities for the use of OIS. A discussion of the uses of each of the reports follows.

The Workgroup Summary Reports   The Workgroup Summary Reports provide the basic information which describes a workgroup, its main tasks, links, and its use of technology to support those tasks. These include the Workgroup Report, the Linkage Report, the System Usage Report, the Linkage Comments Report, and the System Comments Report.

The *Workgroup Report* summarizes the data collected by the functional analysis form (FAF). The *Linkage Report* summarizes the links to other workgroups for each task listed in the Workgroup Report. This is the information contained in the TLF. The *System Use Report* summarizes the use of support systems for each task listed in the Workgroup Report. It contains the information from the TAF. This set of reports should go to the individual workgroup and then to the next level of management. They can be used to verify the original data, and provide a record of the raw information. Some of the questions raised by these reports are:

Do the tasks listed fairly represent the activities of this workgroup?

Are the activities that describe these tasks complete and accurate?

Are the support technologies correctly allocated to tasks?

Are comments on the support technologies recorded?

Are the links fairly recorded?

Are there comments regarding the utility of the links or problems with them?

Can the manner in which tasks are distributed in the workgroup be improved?

What opportunities are there for use of technology to further support workgroup tasks?

Are there overall tasks or technologies which have not been captured yet?

While this set of questions is not exhaustive, it does provide guidance as to the types of questions that the researcher and the managers involved should be asking.

Summary Reports   At the next level of management, where the individual manager will have his or her own workgroup report, plus those of immediate subordinates, there are additional questions that must be considered. These include:

Are the resources of the subordinate workgroups appropriately divided?

Are the links between workgroups operating smoothly?

Are there better ways to organize these workgroups to improve better effectiveness and efficiency?

Are there technologies which can be used to facilitate the key tasks of this set of workgroups taken as a whole?

Are we focusing on useful organizational tasks?

In order to answer these questions we use the *Linkage Summary Report*, that lists the names and strengths of the links to other workgroups over all tasks; and the *Summary of System Use Report*, that lists the name of the available systems and the key tasks associated with each. In addition the *Linkage Comments Report*, which lists all the comments made on the links to workgroups, and the *System Comments Report* which lists all the comments made on the systems, provide useful insights. In addition, we use the *Links and Support Report* that categorizes workgroups on their relative number of links and relative use of technology. As we discussed in the example of BEI, four categories of workgroups are identified:

(1) workgroups with high support and a high number of links;
(2) workgroups with high support and low links;
(3) workgroups with low support and high links; and
(4) workgroups with low support and low links.

This set of four categories should provide a catalyst to encourage senior corporate executives to ask questions regarding organizational structure and design. One approach is to use the HIT data to rearrange tasks into different packages and to re-examine the organizational structure. Note that all of the summary reports can be provided at various levels of aggregation ranging from the individual workgroup level, through the departmental level, to a summary of all recorded workgroups.

Providing information to information systems   The *Links and Support Report* provides a general guide to the IS department in allocating manpower and other resources. Those workgroups in the high–high category are critical, since they use existing technology heavily and are closely linked to other workgroups. This implies that a very thorough analysis of these groups must take place before replacing existing systems. It also implies extensive consultation with other workgroups who are linked to these groups. The low–low workgroups are virtually stand-alone groups. This implies that one may be able to solve immediate problems with stand-alone technology and without extensive consultation with other workgroups. The remaining two categories (high–low and low–high) must be considered on a case-by-case basis.

Other summary reports provide answers to such questions as:

In redesigning a particular system, which workgroups should be consulted?
How can we be sure that important inter-workgroup links are maintained?
How can we set priorities for system development?
What are the key problems that are currently being experienced with our existing systems?
How can we accommodate possible future needs?

SUMMARY

In this chapter we have discussed a variety of methodologies for requirements analysis and discussed one technique (HIT) which provides data to help management and workers control the process of requirements determination, system design and implementation. Our view is that organizational members must focus on key tasks and organizational responsibilities, and leave the technical aspects of OIS design and implementation to technical experts. However, this does not mean that management can leave the project management to the technical people. Ideally, management controls the process, employees control the content and OIS/MIS experts control the technology.

BIBLIOGRAPHY

Bailly, J. and Pearson, S. (1983) 'Development of a tool for measuring and analyzing computer user satisfaction', *Management Science*, **29**(5), 530–540.

Benjamin, R. (1971) *Control of the Information Development Cycle*, Wiley Communigraph Series, Wiley-Interscience, New York.

Berrisford, T. and Weatherbe, J. (1979) 'Heuristic Development: a redesign of systems design', *MIS Quarterly*, **3**(1), 11–19.

Bowman, B., Davis, G. and Wetherbe, J. (1983) 'Three stage model of information planning', *Information and Management*, No. 6, pp. 11–25.

Burnstine, D. C. (1980) *BIAIT: An Emerging Management Discipline*, BIAIT International, New York.

Checkland, P. (1981) *Systems Thinking, Systems Practice*, John Wiley & Sons, Chichester.

Conrath, D. W., Higgins, C. A., Irving, R. H. and Thachenkary, C. S. (1983) 'Determining the needs for office automation: methods and results', *Office: Technology and People*, **1**, 275–294.

Damodaran, L. (1981) 'Measure of user acceptability', in B. G. Pearse (ed.), *Health Hazards of VDT's*, Loughborough University of Technology, pp. 61–70.

Davis, G. and Olson, M. (1985) *Management Information Systems; Conceptual Foundations, Structure and Development*, 2nd edn, McGraw-Hill, New York, pp. 573–495.

Davis, K. (1981) '*Human Behavior at Work*', McGraw-Hill, New York.

Dumas, P, du Roure, G., Zanetti, C. Conrath, D. and Mairet, J. (1982) 'MOBILE-Burotique: prospects for the future', in N. Naffah (ed.) *Office Information Systems*, North-Holland, Amsterdam, pp. 471–480.

Eason, K. (1982) 'The process of introducing information technology', *Behavior and Information Technology*, **1**(2), 199–213.

Federico, P. (1980) *Management Information Systems and Organizational Behavior*, Praeger, New York.

Hirschheim, R. A. (1985) '*Office Automation: A Social and Organizational Perspective*, John Wiley & Sons, Chichester.

Higgins, C. A. and Safayeni, F. R. (1983) 'Specification and evaluation of office automation technologies', *Journal of the Office Systems Research Association*, **2**(1), 53–57.

Irving, R. H. (1983) 'A heuristic for implementing office automation systems', *Journal of the Office Systems Research Association*, **2**(1), 57–59.

Kriebel, C. (1979) 'Evaluating the quality of information systems', in N. Szyperski and E. Groucha (eds), *Design and Implementation of Computer-Based Information Systems*, Sijthoff & Noordhoff, The Netherlands, pp. 29–43.

Markus, L. (1983) 'Power, politics and MIS implementation', *Communications of the ACM*, **26**(6), 430–444.

McKeen, J. (1983) 'Successful development strategies for business application systems', *MIS Quarterly*, **7**(3), 47–65.

Pava, C. (1983) *Managing New Office Technology: An Organizational Strategy*, Free Press, New York.

Phyrr, P. A. (1970) 'Zero-base budgeting', *Harvard Business Review* **48**, 111–121.

Raiffa, H. (1970) *Decision Analysis Under Uncertainty*, Addison-Wesley, Boston, MA..

Rockart, J. F. (1979) 'Chief executives define their own data needs', *Harvard Business Review*, March-April, pp. 81–93.

Tapscott, D. (1982) *Office Automation*, Plenum, New York.

Warren, B., Irving, R. H. and Higgins, C. A. (1984) 'An introductory framework for a computer-based information system', Special Report, University of Waterloo, Department of Management Science, University of Waterloo, Waterloo, Ontario, 6 May.

Wetherbe, J. C. and Davis, G. B. (1982) 'Strategic MIS Planning Through Ends/Means Analysis', MIS Research Center Working Paper.

Zani, W. M. (1970) 'Blueprint for MIS', *Harvard Business Review*, November-December, pp. 95–100.

SUPPLEMENT 7.1: WORKGROUP REPORTS

Workgroup report for the executive office

(1) WORKGROUP: Executive Office   DATE: 10 July 1988
Interviewees: Joe Smith
               Harry Jones
               Sam Levine
Account clerks (5): Susan Sharp, Ann Madden, Lorraine Que, Karen Long, Pat Foley

(2) MAIN FUNCTIONS OF WORKGROUP: Manage overall operations of BEI and set strategic directions for future.

(3) KEY TASKS

KEY TASKS	ACTIVITIES
1. Coordinate production, marketing and design departments	Weekly management meetings
2. Develop new strategic thrusts	Survey industry, intelligence reports
3. Develop long, medium, short range plans	Yearly planning meetings
4. Maintain accounting information	A/R, A/P, GL
5. Ensure sound business practices	Audit trails, financial management, credit checks

(4) WORKGROUP MEMBERS

WORKGROUP MEMBERS	MAJOR TASK RESPONSIBILITIES
1. Joe Smith, President	1,2
2. Harry Jones, Comptroller	3,5
3. Sam Levine, Accountant	4,5
4. Accounting clerks (5)	4

TASK LINKS

WORKGROUP: Executive Office   DATE: 10 July 1991

Importance of link

Group	Low	Medium	High	Problems and concerns
Executive Office			1,2,3,4,5	(All) Extensive internal communication
Production department	1,3	4,5		
Marketing department	1,3	4,5		(1, 4) Don't keep us up-to-date
Design department	1	2,4	3,5	(3) Often late with plans
				(4) Sloppy reporting

TECHNOLOGY USE

WORKGROUP: Executive office   DATE: 10 July 1991

Technology	Usefulness of technology			Problems and concerns
	Low	Medium	High	
Telephone			1,4,5	(All) Too many busy calls
Typewriter			1,2,3,4,5	(All) Secretary overloaded
Copier			1,2,3,4,5	
PC–CAD				

Workgroup Report for the Production Department

(1) WORKGROUP: Production department   DATE: 10 July 1991
Interviewees: Adam Buzkowski, production manager
                       Joe Green, maintenance foreman
                       Ken Cook, production scheduler
                       Len Ouchi, lead hand
                       Ron McGregor, production worker (union steward)

(2) MAIN FUNCTIONS of work group: To produce BEST figures and accessories efficiently

(3) KEY TASKS	ACTIVITIES
1. Production scheduling	BEST line, accessories line
2. Maintenance of production line	Regular and emergency maintenance
3. Quality control	Periodic tests, analysis of returns
4. Reporting to executive office	Weekly, monthly, yearly
5. Implementing design changes	New jigs and fixtures
6. General management	Hiring, firing, promotion, job descriptions, etc.

(4) WORKGROUP MEMBERS	MAJOR TASK RESPONSIBILITIES
1. Adam Buzkouski	6,4,1,5
2. Joe Green	2
3. Ken Cook	1
4. Len Ouchi	3,5
5. Production Workers (9)	2,3

TASK LINKS

WORKGROUP: Production department   DATE: 10 July 1991

Group	Low	Medium	High	Problems and concerns
		Importance of link		
Executive office	_____	_____	4	(4) Request too much information
Production department	_____	_____	1,2,3,4,5,6	_____
Marketing department	1	_____	_____	(1) Too many changes
Design department	_____	5	_____	(5) Unrealistic designs

TECHNOLOGY USE

WORKGROUP: Production department   DATE: 10 July 1991

Technology	Low	Medium	High	Problems and concerns
		Usefulness of technology		
Telephone	1	2,4	_____	Always busy
Typewriter	_____	_____	1,2,4	_____
Copier	_____	1,4	_____	_____
PC–CAD	_____	_____	_____	_____

Workgroup report for the marketing department

(1) WORKGROUP: Marketing department   DATE: 10 July 1991

    Interviewees: Pete Dworkin, marketing manager
                   Lenny Penny, salesman (Western Region)
                   Carrie Waters, salesman (Eastern Region)
                   Kim Steele, salesman (Northern Region)
                   Stu Kimball, salesman (Central Region)

(2) MAIN FUNCTIONS OF WORKGROUP: To increase sales and market awareness of BEST and accessories

(3) KEY TASKS

KEY TASKS	ACTIVITIES
1. Maintain existing client base	Phone lists, visits
2. Survey competitive environment	Market seminars, expositions
3. Develop new markets	Prospect lists, etc.
4. Manage sales efforts	Productivity reports, bonuses, incentives

(4) WORKGROUP MEMBERS

WORKGROUP MEMBERS	MAJOR TASK RESPONSIBILITIES
1. Pete Dworkin	4,3,2
2. Lenny Penny	1,3,2
3. Carrie Waters	1,3,2
4. Kim Steele	1,3,2
5. Stu Kimball	1,3,2

TASK LINKS

WORKGROUP: Marketing department   DATE: 10 July 1991

Group	Importance of link			Problems and concerns
	Low	Medium	High	
Executive office		1,3		
Production department		1		(1) Scheduling problems
Marketing department			1,2,3,4	(All) Don't talk enough
Design department	1			(1) Too slow
External–clients			1	
External–other			2,3	(2,3) Government, industry associations, etc.

TECHNOLOGY USE

WORKGROUP: Marketing Department   DATE: 10 July 1991

Technology	Importance of link Usefulness of technology for task			Problems and concerns
	Low	Medium	High	
Telephone			1,2,3,4	Always busy
Typewriter		1,4		(1,4) Slow turnaround
Copier		1,4		
PC–CAD				

Workgroup report for the design department

(1) WORKGROUP: Design department   DATE: 10 July 1991
   Interviewees: Hal Marks, senior designer
                  Rick Sutton, Jr. designer
                  Al Bradey, Jr. designer
                  Cameron J. Stewart, Jr. designer
                  Alex Bundy, Jr. designer
                  Theo Peridis, Jr. designer

(2) MAIN FUNCTIONS OF WORKGROUP: To update and redesign BEST line and accessory line as necessary

(3) KEY TASKS   ACTIVITIES
   1. Design new BEST accessories   Clothing, props
   2. Redesign BEST to meet trends   Follow fashion industry
   3. Supervise and approve designs   Maintain BEST style and quality

(4) WORKGROUP MEMBERS   MAJOR TASK RESPONSIBILITIES
   1. Hal Marks   3,2,1
   2. Junior designers   1,2

TASK LINKS

WORKGROUP: design department   DATE: 10 July 1991

| Group | Importance of link | | | Problems and concerns |
	Low	Medium	High	
Executive office	3	————	————	(3) Little contact
Production department	————	1,2	————	(1,2) Resist changes
Marketing department	————	1	2	————
Design department	1,2,3	————	————	(All) Need more staff

TECHNOLOGY USE

WORKGROUP: Design department   DATE: 10 July 1991

| Technology | Usefulness of technology | | | Problems and concerns |
	Low	Medium	High	
Telephone	1,2,3	————	————	————
Typewriter	————	————	3	Need word processing
Copier	————	————	————	————
PC–CAD	————	————	1,2,3	Need a new machine

Chapter 8
IMPLEMENTING AND EVALUATING OIS

This chapter discusses the problems and issues involved in implementing and evaluating OIS and extends the issues raised in Chapters 6 and 7. Implementation and evaluation are combined because the two processes are closely related. We assume that the needs assessment, target area identification, strategic planning, and systems development and procurement have already been completed. At this point the organization has ordered the hardware and software, and has determined who will use specific systems, and for which task(s).

The main elements involved in the implementation process include management of the change process, development of education and training programs, and management of the physical process of implementation. The evaluation process includes definition of operational performance criteria, data collection and analysis, and assessment and action resulting from these data.

The consequence of the implementation process will be a reconfiguration of the organizational, social and technical environment. Consequences of the evaluation process includes a variety of ongoing adjustments and changes to the technical and organizational systems comprising the OIS. Ideally, evaluation will be an ongoing process. In part the value of this approach is that the organization can take advantage of the double-loop learning that takes place through use of the OIS. Double-loop learning is the process by which the underlying assumptions and theories of a system are recognized versus single-loop in which the operational skills alone are learned (Argyris, 1982).

THE IMPLEMENTATION PROCESS

The implementation process, as we have defined it, begins once the OIS has been completely designed and the equipment has been ordered. Clearly a main focus of implementation is to get the hardware and software installed and running properly. At this level implementation is strictly a technical process. However, to be successful the organizational and individual components of implementation also must be properly addressed. Management of the change process is one of these components, as is the development of education and training programs.

Because all aspects of OIS design and implementation are interlinked it is hard to separate implementation problems from systems planning, needs analysis and design problems. Furthermore, the resources for support and training in the implementation process must be budgeted well in advance to ensure adequate time and money is available once the project is complete. If money is too tight, or there are scheduling difficulties—as is common towards the end of a project, system performance may not be able to achieve expected levels, thereby compromising the investment made in their installation.

Managing the Change Process

The process of implementing an OIS is both a social and a technical phenomenon. Strassman (1985) contends that changes in organizational behavior are often quite slow and, as a consequence, the realization of benefits of office automation or OIS are often paced by how quickly people can change their behaviors. Consequently, when implementing an OIS, it is important for the organization to ensure that the OIS is being inserted into a receptive social environment. Since an OIS usually represents a change from the *status quo*, a climate conducive to change must be created.

Change is often seen as threatening. This may happen for a variety of reasons. It may be fear of the unknown, concern about changes in the work dynamics, or it may be concerns about job loss, job degradation or changes in social contact.

One approach to minimize the negative reactions to change is to involve employees in the design and implementation of the OIS. It is almost a cliché that if people are involved in the change process they will accept it. However, it is an important cliché and one that is discussed more frequently than acted upon.

User involvement produces better information about task processes, and provides a valuable opportunity to get people thinking about the broader context of their jobs. The process of designing and implementing an OIS is an opportunity for the organization as a whole to reflect upon what it is doing, why it is doing it, and how it could do it better.

For example, in one organization we studied, a new computer-based system was to be introduced for handling clients' financial accounts. The new system was modeled on existing systems. One operational manager investigated the proposed system and found that 50% of the clerical workers were in fact reconciling errors made by the rest of the clerical workers, who were taking data from one computer system and entering it into another computer system. No-one had communicated with the workers effectively while developing the proposed design for the new system, and apparently none of the managers had stopped to think about the nature of the work being done. All employees were concentrating on their own narrow job responsibilities. Only one manager had the time, energy and insight required to tie the jobs together. As a result of his actions the proposed OIS was considerably simplified, and the amount of work required to manage the customer accounts substantially reduced.

The use of the HIT technique as discussed in Chapter 7, or the use of similar sociotechnical processes (e.g. ETHICS) provides a mechanism for effective user participation, which helps to avoid replicating redundant procedures or perpetuating inefficient processes. In the absence of such techniques it is imperative that management form an implementation/evaluation committee (IEC) composed of users, management and a technical representative. The latter is *ex officio* and functions as an information source.

In cases where HIT or a similar technique is combined with a participative planning process, the IEC is a natural extension to the needs analysis and planning process. However, even in these cases it is still useful to have the IEC independent of the earlier planning and implementation committees to avoid the possibility of bias. The function of the independent committee is to act as a steering committee for the implementation team and to manage subsequent evaluation efforts. The IEC provides a bridge between those who actually implement the OIS and those who are affected by it. The committee provides an appropriate vehicle to allow the free expression of fears and concerns with the implementation process, and to ensure that legitimate problems are raised and addressed.

From a managerial perspective, implementation involves identification of key individuals or opinion leaders in each group, getting them involved in the process and informing them about what is going to happen and what the likely impacts are. These people can then act as information resources for other workers who may not be as closely involved in the process. In addition, a general awareness program for all employees is a key step. This involves a series of meetings with employees, articles in the corporate newsletter, and information sessions where people are free to air their concerns and expectations.

The argument often made against having these sessions is that 'you'll only stir up the employees or cause anxiety.' It is better, however, to have the anxieties aired and out in the open before the equipment comes in rather than wait until after the system is installed and find that the employees are quietly sabotaging your efforts. Furthermore, the assumption that controlling information will control rumors is naive in the extreme. During a major reclassification of civil service jobs in the Canadian federal government, information was severely restricted. The motto among civil servants during that period was 'If you have not heard a rumour by 10 a.m. start one!' Other typical arguments are found in Table 8.1.

Involving Unions in the Change Process

Unions may use the introduction of an OIS as an opportunity to deal with such issues as job changes, reclassification, or technological obsolescence. The union may try vigorously to control the introduction of technology to protect its

TABLE 8.1   Spurious arguments against user awareness sessions

(1) The users are too busy with their current jobs to be bothered. [*This narrow focus will never allow for process improvement.*]

(2) We cannot afford the cost of work downtime while the awareness sessions are taking place. [*The costs are often minimal as the session times are usually only a few hours and can take place after normal work hours.*]

(3) All users are not affected equally so we would be wasting their time. [*It is easy enough to arrange different information sessions targeted towards various groups. Also it is beneficial for users to be aware of tasks up and down stream from their own operations.*]

(4) We may have to change our mind and we would look stupid. [*This can be covered by stating up front the plans are temporary. An honest explanation is sufficient to maintain the respect of the staff. Hidden messages and covert discussions usually bring on more concern and distrust.*]

(5) We may have to change our mind and the time spent in awareness so far will be wasted. [*Initial education will at least get users thinking about the processes to be changed and their impacts.*]

(6) Only management knows what is best for the company, these concerns are for too high a level. [*Management are usually coordinators. Information about operational processes is always better at the end-user level.*]

(7) Users do not understand the technology anyway. [*Information sessions should never be presented in a technical format. The object is to get participation and feedback on the impact on job processes, not an evaluation of the technical abilities of the potential system.*]

(8) Information will only give users time to develop counter-arguments and reasons for resistance. [*They will develop these anyway after the system is implemented, and then quietly sabotage the system.*]

(9) User information and awareness sessions will just lengthen the design and implementation time. [*Inappropriate systems and poorly understood implementation procedures lengthen the time even more.*]

(10) It is easier to apologize than to ask permission. [*After too many apologies, management (or IS) will have lost the respect and trust of end-users for a long time.*]

(11) Management (or IS) must maintain control over information for security reasons. [*This is true mainly in highly sensitive defense companies. Normally the users have to know how other areas will be using the systems to ensure their own contribution provides maximum effectiveness.*]

(12) Management (or IS) does not want to scare the users before it has to. [*Acceptance must be built up before implementation or there will be a lag in effective system use due to fear and uncertainty.*]

workers in terms of layoffs and staff reductions. It may also argue vigorously that those people who have been trained in the new technology must be upgraded because their skills have been enhanced.

Insofar as possible it is a good idea to involve unions in the implementation process. In part this is strictly political in the sense that if they know what is coming, and have been invited to participate in the process, they have less of a basis on which to make arguments that management was being unfair or unreasonable down the line when the material is actually installed. Ideally, a union representative would be part of the implementation committee. If this is not possible, at least union representatives should be invited to air grievances and concerns to the committee.

A much more radical proposal for union involvement is presented by Heather Menzies (1989) in her book *Fast Forward and Out of Control* (pp. 245–251). She advocates a process of co-management of technological and organizational change, recommending that this process be written into union contracts. While some die-hard capitalists will have apoplexy at this concept, the results from trials in Europe appear to be encouraging. Not only have innovative ideas for systems design been developed, employee resistance to, and sabotage of, completed systems has been virtually eliminated.

Ergonomics and the Workplace

One important area where employees and unions have been involved is the ergonomics of the workplace. Many unionized offices find that the unions are not tremendously interested in task design, but they are vitally interested in the ergonomics of the situation. This provides a clear technical issue where one can involve the employees. This involvement can include the design of the workstations and physical layout, always, of course, under the proviso that management has control over the costs.

Though the ergonomic issues associated with using OIS are well understood, they are often ignored. One finds examples of terminals placed toward uncurtained east- or west-facing windows so that sunlight completely washes out the terminal screen. Other examples include: excessive noise from printers, poor indoor lighting systems which cause eye strain, inadequate seating systems, improper ventilation and a host of other problems.

Though most people can adapt to physically inappropriate systems if there is enough incentive for using them, their productivity is likely to be impaired. Consequently, management must make sure that there is an adequate plan for physical layout of the equipment, and that the office furniture and environment is designed to support the workers and the equipment properly.

In one office a minicomputer used to go down every time it rained. This was a consequence of inadequate humidity control; when the humidity went up, even moderately, the computer would short out and crash. It cost $10 000 to rectify the problem.

Noise from printers and keyboards can be a problem. Is there a more insane situation than having the productivity of a $90 000-a-year employee reduced for the want of $1000 of sound-reduction material. Clearly sound-absorbing materials should be strategically placed throughout the office. Noise levels should be monitored on a yearly basis at least.

As we mentioned earlier, lighting and eye strain are major ergonomic issues. Certainly no-one should be forced to use a computer terminal that is set against a bright sunlit background. Secondly, some people have eyesight problems;

TABLE 8.2   Examples of ergonomic issues that should be considered

(1) *Proper desks*
  height of desks
  height of keyboard rest
  proximity of terminal to normal work area
  proper place to set papers for entry into terminal
  pre-drilled holes or slots for cables and power cords
  proper room under the desk for comfortable positioning of legs and feet
  positioning *vis-à-vis* doorways and windows (one does not like to be startled by the unexpected entry of a colleague)
  is there adequate work space for left-handed people?

(2) *Proper chairs*
  height of chair
  flexibility in adjusting chair position (for height, and back support)
  casters for smooth movement if files must be retrieved frequently
  additional footstool for comfortable leg positioning
  armrest positioning (so as not to interfere with typing)
  fit of chair under the desk or keyboard tray to be used

(3) *Computer terminals*
  coloring of screens (certain colors are easier on the eyes)
  positioning via windows or provision of antiglare screens
  positioning via doorways for content security
  size of screen versus frequency of use (small laptop or portable machine screens are frequently too small for constant data entry)
  positioning relative to the keyboard and the CPU (both should be a comfortable distance away so it is not necessary to bend and twist to enter data or insert disks)
  layout of all users' keyboards and design of CPUs should be similar enough to allow users to switch machines with little acclimatization

(4) *Lighting*
  level of lighting
  fluorescent lighting can emit high-frequency flickers if not carefully monitored
  lighting should be focused on working papers, not terminal screen
  too much lighting can raise the temperature of the work area unnecessarily
  control of lighting should be within easy reach of the worker

(5) *Level of noise*
  placement of printers relative to work area
  level of fan noise from CPU or other peripheral equipment
  level of noise from air conditioning
  placement of work area vis a vis hallways or other traffic flow

(6) *Decor and general environment*
  terminal cables and power cords should not be trip hazards
  placement of power and LAN or mainframe outlets should be convenient to the machines being used
  there should be a sufficient number of outlets to support all peripheral electrical equipment other than the computer, e.g. for calculators, desk lamps, etc. so the users do not have to bend over and rearrange the cord plugs to use different equipment
  temperature must be controlled, air conditioning should be adjusted every time new equipment is installed
  the environment must remain free of dust and smoke for long equipment life and user comfort
  there should be a separate area for equipment maintenance so that normal work processes are not interrupted
  adequate carts with casters for the inevitable moving of sensitive and heavy equipment, or for machinery that must be shared between different users
  furniture and/or work area must provide enough privacy to allow the workers to concentrate on their tasks without unnecessary distraction from other conversations or movement.

for example, those who wear bifocals may need special terminal screens in order to accommodate their vision. Examples of other ergonomic issues to be considered can be found in Table 8.2.

In any event, management must ensure that consideration of ergonomic issues is part of the planning process. In developing an ergonomics plan extensive consultations with individual workers and with union representatives is wise. We recommend a joint union/management task force to address ergonomic issues.

CHANGING JOB ROLES

It is clear that with the introduction of office automation systems the roles and functioning of the jobs in the organization will change. Management personnel have a number of alternatives. They can simply lay off excess workers, increase the number of part-time workers, or develop a displaced worker pool that would be available for a variety of jobs throughout the organization. On the other hand, they can retrain people for the same functions but on a computer, or retrain people for enhanced functions which include greater variety or more difficult tasks than previously. A final option is complete job redesign.

Job enhancement is usually superior from an employee perspective. Serious problems develop when in fact the jobs are degraded rather than enhanced; where, for example, an employee is moved from a very interesting job with variety in it to a very boring restrictive job with no variety whatsoever. Unfortunately, early office automation implementations involved job degradation. Typing pools were introduced, production lines were set up in the office for many of the clerical workers, and much of the social interaction and satisfaction associated with doing the jobs was removed.

Management must be very concerned with job redesign and must take an active role in the design of the jobs. Managers *must not* abdicate responsibility for the job design; in effect leaving it to systems analysts and technical people who are designing the office information system.

Associated with the job design and changing job roles are a number of social issues. One important issue is the amount and nature of social contact that is required or available in the job. For example, if one has extensive customer contact, managers must assess whether or not they are rewarding employees for taking time to deal well with customers or are rewarding them for handling as many customers as possible.

Too often, this social contact is seen by managers as a 'soft' issue, or something with negative implications in the sense that the social contact is seen as unproductive time. These managers would radically change their attitude if they talked to employees in organizations with extensive computerized systems already in place. In one organization contacted recently, it became apparent not only that the employees in different departments did not know what other departments were doing, but that employees within the same departments did not know what their colleagues were doing. This lack of understanding of the job pressures of co-workers was a direct consequence of a computer system that kept the employees at their desks doing their own narrow bit of work, and did not provide sufficient opportunity for interaction.

Serious consideration of the social contact issue may indicate that:

more time should be spent between employees to share information and ideas; and

that perhaps more time should be spent with customers.

In other words, the time saved by the system in terms of routine or tedious work that the computers are now doing could perhaps best be spent in social contact. There also may be a need to promote other (non-job) social interaction to compensate for the loss of social contact in some other jobs. For example, the company may decide that it is important to have more picnics, dances, contests, company-sponsored recreational activities, or other events to promote social contact among workers to build more of a team spirit.

The importance of human contact and commitment is highlighted by Zuboff (1982). She discusses two credit-collection offices in the same company with the same system. In one there was a 'sweatshop' environment with stringent quotas, stringent supervision and very little social interaction. In this office there was extensive system sabotage, a 100% turnover rate per year and an extremely negative environment. In the other office there was a more adult approach, where the supervisors acted as resource people. In other words, a supervisor used the information generated by the system to act as coach to the people, as opposed to a judge. Individual workers were permitted to chat as long as they were not being unreasonable about it, social interaction was encouraged, as was productivity, and people were treated as adults who were able to manage their affairs. The collection rate in the office was $250 000 higher than in the other office, the information provided was better than in the other office, and the turnover was extremely low. The conclusion that we can draw from this is that the social context or social matrix in which the system is embedded is an important determinant of the success of the OIS.

In addition to the social matrix in which the OIS is embedded, there is also a matrix of power relationships in the organization. New communication channels may limit or increase access to information, and as a consequence may change the resulting formal and informal power structure within the organization. It may give rise to excessive security concerns, or it may create resistance to change from those people who could be losing control over the information.

A good description of the effects of changes in this power matrix which underlies the implementation of an information system is given by Markus

(1983). Markus found that the reasons for system use or non-use were in fact a complex mix of technical, social and power-related reasons which went far beyond any strictly technical analysis. It is very clear that if a new system changes control over information and decision making, those groups which currently have control over those factors will likely resist the system. Similarly, it is likely that if the system gives excessive control over decision making and information to a particular group, this group will likely be in favour of the system.

EDUCATION AND TRAINING

Prior to installation, management must run education sessions to inform the employees about the new system, what is going to happen to them, when it is going in, what kind of skills they are going to learn, and what kind of resources management is going to provide to ensure they learn the skills. These sessions should substantially reduce employee uncertainty. The other benefit, of course, will be that understanding of system capability should lead to better utilization of the system once it is installed.

Managers must be careful not to underestimate the value of user feedback and involvement. It is extremely likely that during these sessions some of the employees will identify a number of issues that had been forgotten, particularly if the employees were not properly consulted during the design of the system. The education session should be broad in scope and should include a description of the purpose behind the change. In addition, it should identify changes in workflows and procedures; indicate how the jobs and people will be rebalanced or reorganized to fit into the new office structure; and identify benefits and expected problems that will come with the system. Session leaders must be prepared to answer questions and to consider user feedback. Ideally, the session would involve some hands-on system use so people can touch the system and play with it. A demonstration laboratory where people could wander in and play with the system at their convenience is a good idea.

Training

There are some general issues surrounding training that are important for managers to understand. First of all, training requirements are often grossly underestimated and are often cut back as a cost-saving measure. Most competent consultants and researchers in the field of office information systems will tell you that your training budget should be as much as 50% of your technical budget. In other words, if you are installing a system worth $100 000 you should be prepared to spend $50 000 training your employees.

Learning a new computer system has often been compared to learning a musical instrument. For an individual it takes considerable time, and is often a frustrating process until a certain level of skill has been obtained. Consequently, training programs must be carefully designed. The design should consider both the number of skills necessary and the level of skill required.

Training must have a very clear purpose. Simply providing physical access to equipment is not enough. Access must be thought of as a combination of physical access, access to the system, access to the command system and perception of the access that is based on previous experience with organizational implementation of office information systems. The ideal training system must be able to allow for individual differences between fast and slow learners, for different interests and different levels of need, and for the varying previous experience and aptitude in dealing with computers that various employees will bring to the training session.

Ideally, the training design would be tested to see if it provides training in specific subskills. It should also provide training in the total package of the macro skills needed to operate the system. The level of training provided should be compared to some specific criteria such as:

How much knowledge or experience is actually needed?
How much knowledge is really required to operate the system?

This knowledge must be defined and categorized by job tasks, roles and departmental boundaries. Training must be carefully planned and scheduled, and it should include all users.

One problem with training is highlighted by the introduction of new telephone systems into an organization. In many cases where new telephone systems are introduced, there is an inadequate level of training. Certainly training sessions are scheduled; but in a number of organizations these sessions are very brief, and are not convenient for the users. As a result many users miss them.

The results of inadequate training for use of the phone system may not appear readily, but they can be quite severe. Who knows what the consequences are to an organization when a customer calls and the phone rings nine times? Who knows how customers react when they reach the wrong department and the employee reached in error does not know how to transfer them to the proper department? It is not something one can readily see on the bottom line, but as a manager one must be concerned about these issues.

Another aspect of training should be the gradual introduction of system capabilities. The rate at which people can learn and feel comfortable with computer system features varies; however, it is clear that the amount that any one person can absorb is somewhat limited. It is often best to break the training into chunks, perhaps starting with training on how to use electronic mail first, then going on to word processing or spreadsheets or some other function. Training is best done one-on-one with hands-on sessions and a considerable amount of follow-up and assistance which addresses individual needs. It is important that the employees have an opportunity to air the problems they are finding with it, and to use feedback to improve their training. Not only must you have a training session when the system is being introduced, you need ongoing training as new people enter the company or job rotation moves people into jobs where they have to use aspects of the system for which they were not previously trained. One must also remember that the skills learned at any session will be forgotten if they are used infrequently. Consequently, there should be an opportunity for follow-up training and the system should have an interactive training or help facility that is available to the user at all times. By supplying these help systems, and good reference material and access to other users, the users can learn how to learn. To the extent that it is possible, users should be encouraged to experiment and figure things out on their own.

One of the choices with which managers are faced is a choice between supplier-based training versus in-house training. It is often best to develop some in-house talent. First, supplier training is often inadequate; suppliers view it as a cost, and often give very brief and inadequate training sessions. Second, in-house people with expertise in training provide a corporate resource, available when needed. If at least three or four people in the organization really understand the system it will be a major advantage later on if problems begin to arise. Training is often difficult to get out-of-house for any but the most standard applications. Coordination of ongoing support and training is also better done in-house and is often more cost-effective. Finally, many systems are tailored to a particular organization's needs and require someone with specialized organizational knowledge to be able to effectively train the proper users at the proper time and level.

It should be noted, however, that there are many excellent organizations that train users on industry standard packages (such as Wordperfect, Wordstar, Lotus, Excel, etc.). Because these packages are used by many people, these external organizations can develop the expertise required for comprehensive training programs, and can provide this service at relatively short notice, whenever the training is required for new employees, or refresher courses for older employees.

A related issue that managers must address is the provision of manuals and documentation to the users, both in the corporate library and for the office information systems group. Managers must remember that it may be necessary to supplement vendor materials, and that such things as widely distributed handy reference cards etc., are good learning tools.

Employees must also be trained in how to handle their new roles and adjust to new system skills. For example, Zuboff (1982) discusses a group of credit analysts who had many of their routine functions mechanized on the assumption they would spend the time saved in more professional duties. However, the employees felt that the functions that were mechanized were precisely where their expertise lay, so they did not use the system. What management should have done in that case was to promote the advantages of the system, train employees on the new task required, and adjust the organization so that analysts with more complex skills could advance within the organization. Simply put, it is not enough to assume that people will spend their time in more productive activity, unless guidelines are provided for people on how to use their time more productively. For example, a stock market analyst will not spend time developing more sophisticated models of stock market performance unless he or she has the technical skill to develop those models in the first place.

Carroll (1984) has taken the minimalist approach to training with the assumption that less is more. The idea is that learners will achieve more if they have less to read and less overhead to get tangled up in. Unfortunately, trainers and documenters often feel that their job is not complete until they have documented every aspect and nuance of the system. This level of documentation is appropriate for the systems analysts and support people who must keep the system running. It is not appropriate for people who are learning how to use the system features. Furthermore, people bring their own attitudes and perceptions to the use of any system. If it is not clear how the system works, people will develop their own explanation or theories for how the system works—ones which may be wildly inadequate or inaccurate, but which serve their need to understand the system. A typical response to many training programs is: 'I want to do real work, not just learn how to do everything.' People often skip large sections of manuals on the grounds that 'this is just information.' One of the problems is that many of the manuals assume that the student follows properly through step by step and only give a few explanations or methods of recovering if you make a mistake. A solution is to cut the number of words in the manual, keeping it very simple. This forces coordination of the manual and the training. Another solution is to implement online documentation to help users at the point where they run into a particular difficulty. If help is as easy as pushing another button, it is much more likely to be used. Sometimes it is even wise to omit some items purposely, so that the user must use the system to find out things. Furthermore, the training must anticipate every possible error, or at least the most common ones, and include specific error recovery methods. For example, if you make a particular mistake, then you should do a particular operation to get out of it. The training should focus on real activities with open-ended exercises at the end of each chapter to help the student learn how to use the system. Furthermore, the training should let the learner lead; it should encourage the learner to explore the system by leaving some parts deliberately incomplete.

Managing the Physical Implementation Process

The conceptual part of the implementation process should be gradual. It should allow enough time for adequate individual adjustments and feedback yet not lose the momentum gained through recent systems research and development, and user awareness and education. Users should be able to acclimatize themselves to the system in such a way that they are not overwhelmed by new information and new activities that they have to perform. Training can be performed in an iterative fashion to allow users to integrate what they know into their jobs, and then be trained at the next level of, or more complex, job requirements.

On the other hand, the physical implementation process should be managed using proven project management techniques. Discussion of project management is beyond the scope of this book; however there are many excellent books (e.g. Meredith and Mantel, 1989) available on project management, as well as many seminars open to the general public. Briefly, managing the physical implementation process involves developing a physical plan for the location of the equipment, an installation plan for the timing of installation and a testing plan to ensure that once the equipment is installed it works correctly.

This plan must be coordinated with user workloads and key tasks. For example, if there is a monthly reporting cycle for executive reports, then it would be prudent for a system that is to service these reports to be tested over a period of at least one month, not over one week in the middle of the month. However, installation should be completed in a mid-month week so as not to disrupt the staff when they are at their busiest.

Pilot Projects

For many organizations, developing a complete OIS is too risky. One alternative to full-scale system implementation is the use of a pilot project (see Chapter 6). This provides an opportunity to evaluate a system, and to introduce changes before it is fully installed. Furthermore, management can assess the effect on individual work groups and interpersonal relations, and can identify changes in goals and objectives that seem to be a consequence of the system. A pilot project provides better cost–benefit data, makes the implementation more user-driven, and reduces the risk. On the other hand, it increases the cost and the amount of time required to get a fully operational system implemented.

During the pilot project one can collect data on the implementation process itself in terms of planning, training, participation, user reactions and the technical performance of the system. This is often vital information if a very large-scale implementation is under way. It does provide the opportunity to head off a number of problems in these aforementioned areas.

The pilot group should be carefully chosen. Not only should they be representative of the mainstream of business activities, but they should also have a real problem that technology can solve. In these cases, if the technology works, the pilot users will be enthusiastic supporters of the system. The pilot trial should also test the need for support. You should choose a group whose abilities are more or less representative of other employees in the corporation, as opposed to choosing a group with high technical skills. In this way one can model the training and support requirements for a full-scale system. Furthermore, a pilot test should involve specific evaluation criteria, should be adequately funded for proper testing and evaluation and should not be a series of isolated experiments but a part of the organizational learning process.

Poor pilot trials occur when too theoretical an approach is used and actual business processes are ignored. For example, one E-Mail system was implemented using randomly chosen participants. The pilot failed since the group of randomly chosen employees had no need to exchange information. In this instance a set of interlocked workgroups would have been an appropriate group for the trial.

One organizational learning model is derived from research of Chris Argyris (1982). He terms the learning process that occurs within the context of an existing theory 'single loop-learning.' Learning which directs attention to the underlying assumptions and theories being used is termed 'double-loop learning.' Single-loop learning teaches operational skills; double-loop learning promotes the development of better assumptions and theories. Training for OIS should promote both types of learning. One approach is to encourage employees at all levels to question fundamental assumptions. This is aided by exercises in which employees attempt to identify current theories and assumptions. The explicit identification of existing organizational theories is often enough to lead to improvements. It is best if an OIS can be introduced in a manner that produces an incremental refinement of existing procedures while computerizing them. Subsequently, employees learn new behaviors and tasks that the system will support, and incorporate these into their work roles. This is the basis of what Tapscott, Henderson and Greenberg (1985, p. 78) refer to as the reinvestment model. In this model time saved in doing old tasks in new ways is 'reinvested' in discovering new ways to do new tasks.

As an example, it is important that the secretarial staff can grasp the concept that all computer files can be read, edited and printed. Word processing documents are just computer files formatted to be read by a specific package; the only difference from other packages being the keystrokes used in the data manipulation. It also helps if they understand that the reason for using a word processing package is that it facilitates arranging textual characters in any format desired. This understanding helps the secretary to learn and use other OIS packages that serve different needs better. For instance, a spreadsheet package facilitates calculation of whole rows and columns of data, and allows easy and fast movement of the data to different areas of the spreadsheet. If they understand these basic design concepts they (the secretaries) will not make conceptual errors such as trying to manually add numerical data and enter it into a word processing package.

THE EVALUATION PROCESS

It is difficult to talk about evaluation in an abstract sense since, ideally, an evaluation is built into the very design of the OIS itself. However, there are some actions managers can take even in the event that evaluation was not considered part of the information system specification process. The most important aspect of evaluation is determining whether or not the OIS has had a positive effect on the company. In other words, does it help the organization to function better, and in a more cost-effective manner?

In order to answer the preceding questions there are three aspects to the evaluation that must be addressed: technical, structural and environmental. The technical aspect focuses on the technical functioning of the system. It is dealt with by testing the OIS after implementation to see if the system performs according to the technical specifications that were developed early in the game. The structural aspect deals with the organizational impact of the system in terms of the way it changes workflows within the organization; its effects on user attitudes and social interaction; and its overall effects on organizational functioning and structure. The environmental aspect of evaluation deals with the changes in the organizational environment. These could include extensive environmental scanning to determine if the environmental conditions that pertained when the system was designed are still relevant.

Typically, the process of designing and implementing an office information system will take anywhere from one to two years. Depending on the complexity of the system it may take longer. Given this time-frame it is not inconceivable that the competitive environment in which the organization finds itself may change during the period from when the system was first designed until it has been totally installed and implemented. Therefore, the environmental assessment of the context of the OIS is an ongoing process. Management must, as part of its responsibility, assess the reasons why the system was initially installed and whether or not the environment or the context in which the organization finds itself continues to support those reasons. Many of the reasons for originally implementing the system may have changed. Consequently, it may be that the system itself should be changed to meet the changed circumstances.

As an example, if the organizational strategy for a life insurance company was to use information systems to become a low-cost provider, and everybody else became a low-cost provider, the strategy must change. This might mean that the system needs updating to allow the company to provide a highly differentiated product or to fill some particular market niche that would be more profitable.

Rather than rely on a *post-hoc* role in managing the implementation process, management should take a more proactive approach. This requires their early involvement in the design and specification of the system. In particular, management must play a leading role in the definition of performance criteria, since these will drive the system design process and provide the benchmark for system evaluation.

Definition of Performance Criteria

An important management issue is to ensure that the criteria for system assessment are appropriate and are adhered to. The authors have seen projects where long-term OIS implementations were finally evaluated in terms of what was accomplished, rather than in terms of what was intended. This undercuts the whole purpose of evaluation. There are three major areas requiring the committee's (or management's) attention; the technical assessment, the organizational assessment, and the assessment of the implementation process itself.

Technical Assessment

The technical performance criteria are a set of specifications which, if met, will ensure that the technical components of the system will function properly. These specifications include such items as response time under various load characteristics, reliability, and system capacity. In general this is an area where considerable technical expertise is required. As mentioned previously, it is imperative that these specifications be determined before system design begins. During the evaluation phase the original specifications are systematically tested to ensure that the system is satisfactory.

The most important role for management is to ensure that the original specifications are *independently* tested; that is, the system is tested by professionals who are independent of the design and implementation team. Part of the committee's job is to review the plans for the technical assessment and to review the results from this assessment.

Organizational Assessment

The organizational performance criteria are derived from initial justification for the OIS. Usually these criteria indicate levels of productivity, cost-avoidance or staff attrition that will pay for the system. In our experience these criteria are frequently ignored. The IEC's role is to ensure that periodic assessments of organizational productivity and effectiveness are taken to ensure that the system is delivering on its promises. Furthermore, the IEC must, in concert with senior management, assess the potential strategic impact of the installed system. We recommend a preliminary assessment at 6–12 months and a final assessment at 18–24 months. In this way both short-term and long-term problems can be identified.

Implementation Assessment

While managers may not have the skills to control all aspects of the implementation, they can certainly record problems and conduct an evaluation of the implementation process. This effort should be regarded as an organizational learning opportunity. By assessing the implementation the organization can learn how to do it better next time.

SUMMARY

In this chapter we presented managerial issues associated with the implementation and evaluation process. These concerns are in turn derived from the needs analysis process discussed in Chapter 7, which itself is strongly linked to the underlying planning process discussed in Chapter 6.

At this point we have covered the theoretical issues associated with OIS (Part I) and the operational issues (Part II). In Part III we discuss issues which are developing, and which will occupy management attention over the next few years.

BIBLIOGRAPHY

Argyris, C. (1982) 'Organizational learning and management information systems', *Database*, Winter–Spring, pp. 3–11.

Carroll, J. M. (1984) 'Minimalist training', *Datamation*, 1 November, pp. 125–136.

Culnan, M. J. (1984) 'The dimensions of accessibility to online information: implications for implementing office information systems', *ACM Transactions on Office Information Systems*, **2**(2), 141–150.

Irving, R. (1982) *Planning for Office Automation*, AEL Microtel Ltd, Brockville, Ontario..

Irving, R. and Munro, S. (1983) 'Facilitating the adoption of office technology', ASAC 1983 Conference, UBC.

Kozar, K. A. (1989) *Humanized Information Systems Analysis and Design*, McGraw-Hill, New York.

Markus, M. L. (1983) 'Power, politics and MIS implementation', *Communications of the ACM*, **26**(6), 430–444.

Menzies, H. (1989) *Fast Forward and Out of Control*, Macmillan of Canada, Toronto.

Meredith, J. R. and Mantel, S. J. Jr. (1989) *Project Management: A Managerial Approach*, John Wiley & Sons, New York.

Moder, J. and Phillips, C. (1970) *Project Management with PERT and CPM*, 2nd edn, Van Nostrand Reinholt, New York.

Strassmann, P. A. (1985) *Information Payoff*, Free Press, New York.

Tapscott, D., Henderson, D. and Greenberg, M. (1985) *Planning for Integrated Office Systems*, Holt, Rinehart and Winston, Toronto.

Zuboff, S. (1982) 'Computer mediated work: the emerging managerial challenge', *Office: Technology and People*, No. 1, pp. 237–243.

Part III
DEVELOPING ISSUES IN OIS

In this part we present three developing issues: computerized performance monitoring, flexible work, and end-user computing. These issues have existed sufficently long that a substantial body of research exists from which we can draw inferences for managers. These three issues are major challenges for managers in the 1990s and will continue to be so into the twenty-first century. In each of the three chapters which follow we define the issue, and then present the research and managerial implications which flow from that research.

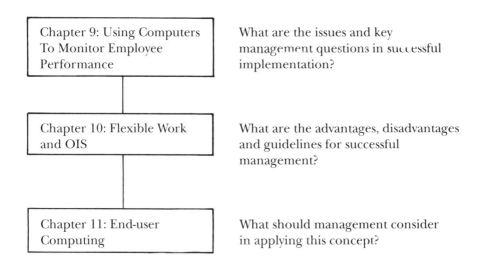

Chapter 9: Using Computers To Monitor Employee Performance	What are the issues and key management questions in successful implementation?
Chapter 10: Flexible Work and OIS	What are the advantages, disadvantages and guidelines for successful management?
Chapter 11: End-user Computing	What should management consider in applying this concept?

Chapter 9

USING COMPUTERS TO MONITOR EMPLOYEE PERFORMANCE

Electronic monitoring represents a complex intervention in the workplace. These systems are inanimate, and yet are often perceived as supervisors in their own right. They are considered to be objective and unbiased; but many employees discover that they can neither explain their performance nor negotiate the consequences of non-standard performance. Computer monitoring systems are, according to some, the domain of managers who believe that 'people do what's inspected, not what's expected.'

Increasing computerization of office functions is providing managers with new opportunities to monitor and control their employees. Using simple computer technology, organizations are gathering detailed, accurate and instantaneous information on employees who work on computer terminals. But the use of computerized monitoring has sparked a heated debate. Many organizations support electronic monitoring because of the improved control and productivity gains that result. Unions, on the other hand, are strongly opposed. They cite factors such as the mistreatment of employees, the return of piece-rate pay schemes and the invasion of employee privacy.

Here we examine management issues associated with computer monitoring. The content is based on three studies of computerized monitoring and on interviews with a wide range of experts. In our first study (Irving, Higgins and Safayeni, 1985), we interviewed 163 clerical employees and their managers across five different organizations. The second study (Higgins, Irving and Grant, 1987a) involved 93 monitored claim processors who participated in a 30–45-minute interview and completed a 110-item questionnaire. Finally, we conducted a national survey of the impact of computerized monitoring in the service sector (Higgins, Irving and Grant, 1987a).

Many white-collar workers are unconcerned about electronic surveillance because they feel it will never affect them. Nothing could be further from the truth. First-line managers are at risk since their supervisory roles may be usurped by electronic performance measurement systems. Furthermore, with the inexorable spread of automation at the white-collar workstation, the extent of monitoring is sure to broaden. Consider electronic time-management systems which allow managers to look at their employees' schedules and book appointments. Is this not a form of monitoring and control? Similarly, recent developments in computerized telephone technology allow management to monitor telephone use on a per-call basis. These systems even permit audio monitoring, where managers listen in on employees' conversations, allegedly to assess the quality of client/employee interaction.

In whatever form, computerized performance monitoring and control systems (CPMCS) are here to stay. Consequently, managers must be aware of the issues resulting from the use of computerized monitoring; be cognizant of the questions that must be answered before deciding to use CPMCS; and be aware of appropriate actions to increase the probability that these systems will be used wisely. In the following sections we address each of these three areas.

ISSUES RAISED BY COMPUTERIZED PERFORMANCE MONITORING

In this section we examine four key issues raised by the existence of computerized performance monitoring in the workplace.

Issue 1: Computerized Monitoring is a Growing Aspect of Work Life

There are no accurate figures on the extent of computerized monitoring. According to the Office of Technology Assessment, from 20% to 35% of all clerical workers in the US are monitored by computer. This estimate may be low. Professor Alan Westin of Columbia University surveyed data processing, word processing and customer service operations in 110 organizations and found that 98% used computer monitors (Westin, 1985). The National Institute of Occupational Safety and Health estimates that two-thirds of all VDT users are monitored (Jacobson, 1984). With 40 million current VDT users, computerized monitoring is becoming an organizational fact of life.

Issue 2: Use of Computerized Monitoring Changes Productivity

At American Express a computer measures the time taken to answer and process a client call. The company reports an overall productivity rise of 5% a year. A similar system is used by Air Canada to rate the productivity of their reservations clerks, and gains of 4% to 5% have been reported since monitoring was introduced.

Reports from other organizations are larger. A monitoring system at Equitable Life gauges employee productivity based on the number of claims processed and the difficulty of each claim. These figures are used to grant salary increases. Substantial productivity gains resulting from the program have allowed Equitable to process 20% more claims with 25–30% fewer people (*Dun's Business Month*, January 1984). The Third National Bank of Nashville found that productivity increased 55% after a monitoring system, together with a pay-for-performance program, was implemented (*Dun's Business Month*, January 1984).

Note that all of the preceding examples reflect increases in volume. In our own studies, and in studies reported elsewhere, strong concerns are raised about such issues as quality of output and the quality of work life. Increasing volume by 50% may not be so impressive if turnover doubles and quality decreases.

Issue 3: Computerized Monitoring Affects Employee Evaluations

Many organizations which use computer monitoring argue that these systems make the performance evaluation fairer because the productivity figures reported by the computer are accurate (the employee cannot falsify them) and unbiased (they are not affected by whether the boss likes the employee).

These claims also have empirical support. In our research (Irving, Higgins and Safayeni, 1985, 1986) we found that a majority of the respondents perceived an increase in the appropriateness, accuracy and completeness of the performance information collected and, as several respondents noted, 'The system doesn't care if you are a friend of the supervisor or not.' When we studied the responses more carefully, we found that those who felt the evaluations were fairer also commented that the monitoring system captured all important aspects of their work. Those who felt computerized monitoring led to unfair evaluations commented that the system only measured volume and failed to capture such important aspects of performance as quality, and customer relations. All employees commented that it is not so much what and how the system collects the data, but what managers do with the data.

Issue 4: Computerized Monitoring Changes Management Style

Organizations also make a strong case for computer monitoring systems as an effective management tool. For example, one system marketed by Clyde Digital Systems in Utah allows a supervisor to watch, in real-time, an employee's interactions with a video display terminal. Another of their systems allows supervisors to record, and subsequently review, an employee's working session on the computer. Although the systems are primarily marketed as security devices to prevent computer fraud, they also serve as management tools. For example, if a supervisor notes that an employee is using a terminal inefficiently or incorrectly, the mistake can be quickly recognized and corrected. Clyde Digital readily admit that 50% of their customers buy the systems to monitor productivity.

On the other hand, merely adopting computerized monitoring will not automatically solve an organization's productivity problems. Electronic surveillance systems can fall far short of expectations, and can create substantial problems if they are not thoughtfully applied. One major problem is the loss of trust between management and employees. Typically this happens when computer monitoring systems are installed with little or no employee input and where the output is used punitively.

KEY MANAGEMENT QUESTIONS

The four management issues raised in the preceding section lead to five questions which must be answered by management *before* deciding to implement computerized monitoring systems.

Question 1: Does the Current System Measure the Right Performance?

Kerr (1982) wrote of the folly of rewarding A, while hoping for B. This is a common mistake that organizations make when they implement computerized monitoring. For example, in one insurance organization we studied, computerized monitoring was used to establish employee performance ratings based on the count and difficulty of claims processed. With an objective measurement the managers in this organization believed they could reward productive employees, encourage slow ones, and better control the work process. Since the monitoring system checked the time taken to process the claim, but not the accuracy of the claim, the processors worked with the attitude 'When in doubt, pay out' (Irving, Higgins and Safayeni, 1986). The hidden costs of overpaying many claims under this monitoring system were staggering.

Another example clearly illustrates the problem of not measuring the right thing. A large US-based insurance company designed a system to measure volume of claims processed. It was implemented together with a daily performance quota. To achieve their quotas, claims processors often put aside their 'scrapwork.' These were claims that, in the terminology of assembly line work, did not fit. Only when a clerk was ahead of her quota would this 'scrapwork' get done (Higgins, Irving and Grant, 1987a). In this case the system was not measuring the right thing. Complexity was excluded and, as a result, complex claims were ignored. Clearly, management must consider which key aspects of performance are critical to task success before deciding on computerized performance monitoring as an option.

Question 2: Do Employees Understand the System?

Many organizations design computerized monitoring systems which are so complex that employees cannot understand them. For example, work measurement experts establish average work times for each of an employee's tasks. These estimates are then used in conjunction with an employee's workload to produce a weighted overall productivity score. These scores may reflect individual productivity in a statistically unbiased way. However, it is difficult for individual employees to relate their activities to these scores.

Equitable Life installed such a monitoring system to count the transactions completed by claims approvers. These workers were then paid wage incentives based on the counts. While the employees liked the wage increases made possible by computer counting, they argued that they did not understand the method used by the company for scoring, and they questioned its fairness. The result: computer monitoring became a major issue in collective bargaining talks between the company and its employees (*Dun's Business Month*, January 1984, p. 38).

Question 3: Does Monitoring Increase Employee Stress?

Two major complaints against computer monitoring are increased stress and potential health hazards. A study by the National Institute of Occupational Safety and Health (NIOSH) showed that heavily monitored clerical workers were under more stress than unmonitored workers (*Potential Health Hazards of VDTs*, 1981). Increased stress as a consequence of electronic monitoring was also evident in our study. Over 70% of the respondents reported a significant increase in stress when their organization implemented a computerized monitoring system (Irving, Higgins and Safayeni, 1986).

There are two sources of stress. First, many organizations which use electronic monitoring subsequently introduce quotas and pay-for-performance programs. These programs tie an employee rate of pay to predetermined production rates. Second, organizations often post individual performance data, which results in excessive competition between employees. Many organizations argue that posting the data results in high productivity because employees do not like to be at the bottom of the list. What frequently occurs is goldbricking, where employees stick to a certain level of productivity despite management's efforts to encourage increases in productivity. We observed several instances of this informal practice.

The most stressful form of electronic monitoring occurs when the system actually paces the work. These systems, such as Citibank's customer sensor, know when an employee is free and thus ready for new work. With systems which pace the work, stress-related health complications often occur. In a study of 2150 American workers, financed by the NIOSH, it was found that machine-paced assembly line workers were 70–200% more likely to develop heart disease than their supervisors (*Dun's Business Month*, January 1984).

Question 4: Will Employees Complain of Mistreatment?

Workers' fears of unfair evaluations and mistreatment are a critical issue for organizations. This is also an opportunity for organized labor. Unions have already taken a strong stand against computerized monitoring. Karen Ringden of the American Civil Liberties Union is spearheading a coalition to lobby against the practice. The list of participants (e.g. UAW, CWA, FAST, 9 to 5, Newspaper Guild, AFFCME) gives us an indication of labor's strong stance.

Karen Nussbaum, a former Harvard Business School secretary and founder of 9 to 5, a national association of working women, is critical of computerized monitoring. In a recent report, she outlined several areas of concern (Nussbaum, 1986). These included mistreatment of employees, invasion of privacy, increased stress, health problems, unfair performance evaluations and the return of quota systems.

Union leaders view electronic monitoring as an opportunity to reverse the tide of decreasing membership by organizing the growing service sector. Campaigns focus on organizing disgruntled office employees working under computer monitoring. As one labor leader stated, 'without clear guidelines for humane and effective use of such systems, management is merely creating good union members' (Higgins, Irving and Grant, 1987a).

Question 5: Is Monitoring an Invasion of Privacy?

This is the final and most controversial issue surrounding computerized monitoring. The crux of the issue can be simply stated: where does the legitimate need of business to collect information on the work being done end, and where does the right of employees to personal dignity and privacy begin?

Interestingly, invasion of privacy is a problem of the new, computerized office. Yet our recollection of the traditional pencil-and-paper office is that such offices were more intensely social and more integrated—in short, less private than the automated offices that replaced them. Is it not curious that, in those offices where computer monitoring systems break the social bond and isolate workers from one another, employees would shout 'invasion of privacy'?

This is likely because the type of work being assessed is different. Previously an employee was assessed either purely on results or on a monitored process with the supervisor visibly present. During the process monitoring, the employees could affect a different work process or procedure if they thought it would satisfy the supervisor. With CPMCS, however, work process can be randomly monitored without the knowledge of the employee. A constant lack of process flexibility and the resulting stress from 'always being on one's toes' can lead employees to complain about a new lack of privacy.

STEPS TO SUCCESSFUL IMPLEMENTATION OF CPMCS

Senior managers must be aware that the misapplication of a control system can have costly effects. In an organization that uses computerized monitoring, where pay depends on the ability to meet a computer-monitored quota, employees' adherence to the abstract logic of a control program could become more important than their traditional loyalty to a supervisor or to the company.

In cases like this, organizations can expect workers to 'pay out' and ignore 'scrapwork.'

Surprisingly, given the heated debate on the subject, a comparison of labour demands, employee attitudes, and accepted principles of work measurement and control shows much agreement. These areas of agreement are combined with the results of our own research to develop recommendations for successfully implementing computerized performance monitoring systems. These recommendations, which are summarized in Table 9.1, are discussed in the following sections.

TABLE 9.1 Summary of issues, questions and actions regarding computerized performance monitoring

---

Management issues raised by computerized monitoring
Issue 1: Computerized monitoring is a growing aspect of work life.
Issue 2: Use of computerized monitoring changes productivity.
Issue 3: Computerized monitoring affects employee evaluations.
Issue 4: Computerized monitoring changes management style.

Questions to be answered before deciding to implement computerized monitoring
Question 1: Does the current system measure the right performance?
Question 2: Do employees understand the system?
Question 3: Does monitoring increase employee stress?
Question 4: Will employees complain of mistreatment?
Question 5: Is monitoring an invasion of privacy?

Steps to successful implementation of computerized monitoring
Step 1: Identify the strategic aspects of performance.
Step 2: Monitor the work, not the worker.
Step 3: Add some high touch to the high tech.
Step 4: Review the monitoring system at least once per year.

Step 1: Identify the Strategic Aspects of Performance

Managers, workers and researchers agree that computerized performance monitoring systems will shift the emphasis toward those aspects of performance, such as production counts and error rates, which can easily be measured. To the extent that these simple numeric measures reflect important aspects of job performance this shift in emphasis is appropriate. However, where a job contains important behavioral dimensions that are not easily quantified, computer monitoring is a potential recipe for disaster.

Management must analyze the probable consequences of computerized monitoring, identify those aspects of performance which are of strategic importance, and ensure that these performance dimensions are emphasized. Some organizations are already doing this. At a 'well-known' Life Assurance

Co., an organization with $7.6 billion in assets, both qualitative and quantitative aspects of performance are important. A claims supervisor commented that:

> The data from our computerized monitoring system records only one aspect of employee performance—the number of claims processed. This is important information. However, we are concerned about other aspects of performance such as interpersonal behavior, and quality of work. If we relied only on the computerized data we would have a distorted picture of employee performance.

Concern for the broader aspects of employee performance is in evidence in many other organizations. Lynden Heck, the Executive Director of the Center for Office Technology in Washington, DC, manages an organization which includes Fortune 100 and 500 member companies. When asked about the use of computerized performance monitoring by the member organizations she noted that many members were becoming increasingly concerned with qualitative measures of performance in addition to the quantitative measures.

Step 2: Monitor the Work, Not the Worker

Computerized systems, because of their ability to collect and store individual productivity data, tend to shift attention toward the individual. However, this is not always appropriate. For example, the privacy debate centers on the confrontation between managers' right to know about the work being done versus employees' rights to privacy. The simplest solution is for management to avoid establishing controls that key on personal productivity.

While there is no best method of using productivity data, our experience leads us to conclude that individual productivity data are best kept at the level of the individual worker and his or her immediate supervisor. At this level, corrective action can be taken in the context of detailed knowledge of both the individual and the job. Similarly, at the next level of management, production for each unit rather than by worker is an appropriate level of aggregation. This directs the second-level manager's attention to relative differences between units rather than to individual differences in productivity.

An example of improper monitoring occurred in one of the organizations we surveyed that had initially posted individual productivity data. However, they found that the work environment became so competitive that productivity was impaired. Senior employees would no longer train junior ones because 'it reduced their productivity' and employees who had formerly been pleasant and cooperative were now antagonistic. When they discontinued the practice, the interpersonal problems were reduced and productivity and turnover improved.

A $1.3 billion gas utility located in Ontario, makes extensive use of computerized monitoring. A senior vice-president notes that the successful use of the monitoring system is a consequence of the way the data are used.

At XYZ Gas the emphasis is on performance by regional office, rather than by individual. In addition, senior management emphasize the use of quantitative data as a management tool to identify potential problems rather than as a whip to spur increased individual productivity. One executive sums up the company's approach by noting that the emphasis is on reasonable expectations. She also notes that the company reassesses the performance monitoring systems twice per year to ensure that the measures used reflect the work being done.

Step 3: Add some High Touch to the High Tech

One of the complaints voiced by the respondents in our studies was that they had less contact with their supervisors. This complaint was justified. The supervisors commented that they spent less time on the floor since they had the productivity data on their desk. At one organization, senior management were quick to note this tendency and quick to take action. A senior vice-president maintains that one of the reasons for the success of computerized monitoring was that supervisors were encouraged to use it to identify productivity problems and then to provide individual coaching and training. He firmly believes that managers must get out of their cubicles and spend time on the floor with their subordinates rather than looking at computer counts of production. (The concept of high tech–high touch is described by Naisbitt (1982), pp. 35–42.)

Step 4: Review the Monitoring System At Least Once Per Year

At the Gas Utility, managers review the performance monitoring system twice per year. One supervisor commented that they have made minor alterations every time it was reviewed, and major alterations at least every other time. The main reason for the minor alterations was that the system was not performing as expected; major alterations were made as a result of a changing competitive climate. This example reflects the information collected during our studies. The periodic reviews are necessary for all computerized systems but have particular importance for monitoring systems in all but the most stable of competitive environments.

It is important to change the goals of the performance monitoring system as the goals of the organization change. One must also ensure that the CPMCS does not cause inflexibility in response patterns to new problems. An example of this occurs when a company service changes from receiving and manually entering client claims or customer orders, to one of strictly processing the data for approval and action. Many employees, now judged on a work process that is no longer required, will have to be reoriented toward the desired results of the new process. Employees who currently enter data will now review the data sent by electronically transmitted files. In this way the employees' performance system should change from quantity of claims or orders processed to accuracy and completeness of claims or orders received.

Another area for concern is that employees' needs may change over time. It is important that this does not cause dysfunctional use of the CPMCS through falsified entries or use of the data for employees' own ends. An example of this is if there is a downturn in an employees' economic condition, and an employee is judged on a quantity of data entered for files processed. To make a higher level of pay the employee could duplicate entries or send through many incomplete or inaccurate entries.

IN CONCLUSION

The automation of the factory has taught us about performance monitoring and control systems, and how to manage them. Senior management must ensure that this hard-won knowledge is also applied to electronic monitoring systems. By viewing computer monitoring systems as control systems, something we already understand, we can avoid repeating the mistakes in the automation of the office that were made in the factory. (The Appendix provides an historical context from which to view performance monitoring activities. We recommend it to readers who wish to investigate this topic in more depth.)

Electronic monitoring, by itself, is neither positive nor negative. Where managers use a simplistic approach they may create an environment where social conflict and job stress inhibit productivity, and where key aspects of performance are ignored because they are difficult to quantify. If computerized monitoring is used judiciously, and other aspects of performance are taken into account, electronic monitoring may save overhead, provide more accurate production information, and provide a fairer picture of employees' efforts.

In the end the issue is not computerized monitoring—it is management.

BIBLIOGRAPHY

Higgins, C. A., Irving, R. H. and Grant, R. A. (1987a) 'Impact of computerized performance monitoring and control systems: perceptions of the Canadian service worker', *Final Report to Labour Canada*, 31 March.

Higgins, C. A., Irving, R. H. and Grant, R. A. (1987) Study commissioned by Technology Impact Research Fund, Labour Canada, Ottawa, Canada, April.

Irving, R. H., Higgins, C. A. and Safayeni, F. R. (1985) 'A preliminary study of computerized and non-computerized performance monitoring systems', Report to Department of Communications, Ottawa, Canada, March.

Irving, R. H., Higgins, C. A. and Safayeni, F. R. (1986) 'Computerized performance monitoring systems: use and abuse', *Communications of the ACM*, **29**(8), 794–801.

Jacobson, B. (1984) 'When machines monitor work', *World of Work Report*, **9**(5).

Kerr, S. (1982) 'On the folly of rewarding "A", while hoping for "B"', in D. Hampton (ed.) *Organizational Behavior and the Practice of Management*, Scott, Foresman & Company: Glenview, IL, pp. 474–487.

Naisbitt, J. (1982) Megatrends, Warner Books: New York.
Nussbaum, K. (1986) (9 to 5) 'Computerized monitoring and other dirty tricks', National Association of Working Women Cleveland, OH, April.
Potential Health Hazards of VDTs (1981) National Institute of Occupational Safety and Health, Washington, DC, US Department of Health and Human Services.
Westin, A. (1985) *The Changing Workplace: A Guide to Managing the People, Organization and Regulatory Aspects of Office Technology*, Knowledge Industry Publications, White Plains, NY.

Chapter 10
FLEXIBLE WORK AND OIS

INTRODUCTION

The proliferation of computing and communications technology has created the potential for flexible work environments. In addition to flexible work schedules, employees can work at locations remote from their central office. In some countries (notably Sweden) satellite work centers are being established in the employees' communities. A report of a full-scale experiment in neighbourhood work centers can be found in *Tomorrow's Work in Today's Society*, published by the Swedish Council for Building Research, Stockholm, Sweden, in 1986.

In North America a variety of work alternatives are potentially available including contract work at home, and part-time and full-time telecommuting. Nilles *et al.* (1976) coined the term 'telecommuting' to describe the substitution of telecommunications for travel. This potential for variety of work alternatives, combined with rising energy costs, led Toffler (1981) to predict that remote work would be one of the major trends of the future. In 1982, Nilles estimated that as many as 10 million people could be working at home by 1990. Interestingly, the August 1990 edition of *Compute Magazine* (p. 78) quotes a figure of 10.3 million computers used for work at home in the USA.

Despite the predictions of Toffler and Nilles, telecommuting has not had an explosive growth rate. To date, few organizations have implemented formal programs, though many have informal arrangements with their employees. This situation will change in the 1990s. One major factor motivating a change is the increasing number of females in managerial and professional ranks. In North America, females make up 50% of law students and 40% of MBA students. As these professionals move into the workplace and begin to have children there will be increasing demands for flexible work schedules. Furthermore, 60% of married women with children work outside the home. Consequently, the demands for flexible work will come not only from professional women with children but also from fathers whose spouses work. With these facts in mind, it is clear that managers must be aware of the potential advantages and disadvantages of flexible work, the managerial issues raised by flexible work schedules and the management actions that are required to implement flexible work programs effectively.

We begin the discussion by defining common terms used to describe various aspects of flexible work arrangements. We then focus on telecommuting and discuss the advantages and disadvantages to this form of flexible work. The chapter ends with a discussion of key managerial actions required to manage flexible work arrangements effectively.

DEFINITION OF COMMON TERMS RELATED TO FLEXIBLE WORK SCHEDULES

In this section we present definitions for terms commonly associated with flexible work arrangements. (We do not discuss flex-time and job sharing in depth here. Those who want more information on these topics should read Ronen, 1984.)

Flex-time

Flex-time normally refers to the practice of allowing employees flexibility in choosing starting and finishing times for work. Typically there is a core time (e.g. 10 a.m.–3 p.m.), when everyone must be at their desks. In addition there is a total number of hours to be worked each week or month. Within these limits employees can arrange their own work schedules. The various flex-time plans vary in flexibility, but most have these two elements. These plans are touted by major cities to ease traffic congestion. In addition they are particularly suited to dual-career families with small children.

Job Sharing

Job sharing is a practice where two people share one job. Time on the job may be divided equally or not, depending on job skills and individual availability. Job sharing has been used to provide a wider base of employment and to maintain the skills of employees who could only work part-time. Today some firms are using job sharing to retain the skills of professionals who have periods of time when they cannot work full-time (e.g. a father or mother with small children).

Remote Work or Telecommuting

Remote work, telecommuting or telework all refer to a process whereby an employee conducts part or all of his or her job from a location other than the employer's normal premises. Normally this implies a heavy reliance on telecommunications software and hardware. The following paragraphs explore the varieties of remote work.

Extended Day Remote Work

Extended day remote work refers to the practice of an employee working at home on a computer in the evenings. This is often combined with the next type of remote work. Professionals and managers make up the bulk of workers in this category and in the next one.

Extended Week Remote Work

Extended week remote work is the practice of employees working at home, using a computer and telecommunications software at the weekends. Frequently this is in lieu of a trip to the office. As mentioned earlier, one finds mainly professionals and managers in this category.

Occasional Remote Work

Occasional remote work is the practice of employees working remotely on rare occasions (e.g. when travelling). Some occasional remote work involves special projects or activities that require a quiet environment (e.g. working on a major project at home). This type of telecommuting is popular among professionals and mid-management.

Part-time Remote Work

Part-time remote work is the practice of an employee working at home part-time. This may be as little as one or two days per week or as much as two or three weeks per month. The point here is that it is a formal part of the job description. This type of telecommuting is found most among professionals such as systems analysts, etc.

Full-time Remote Work

In full-time remote work an employee works full-time at a remote location (home or satellite work center). Employees may come to the office occasionally; however, their normal workplace is their remote location. As noted earlier, there are still few of these formal programs despite the predictions cited earlier. Most frequently, full-time telecommuters are clerical workers; though some computer professionals and consultants telecommute full-time.

Though the five categories described above cover the full spectrum of flexible work, most of the recent studies have focused on full- or part-time telecommuting. While these forms of flexible work are the least common they represent the most potentially disruptive organizational intervention. Consequently, it is important to understand the managerial issues. In the next section the strengths and weaknesses of telecommuting are presented from the viewpoint of both the organization and the individual employee.

ADVANTAGES AND DISADVANTAGES OF TELECOMMUTING

Though there are a large number of articles on telecommuting, many are reports of a few empirical studies or are completely conceptual in nature. In the key articles we found that the advantages and disadvantages of telecommuting were generally characterized from the viewpoint of either the organization or the employee; rarely were both perspectives contained in one article. A complete summary of the reviewed literature is provided in Tables 10.1 and 10.2.

Table 10.1 presents the advantages and disadvantages of telecommuting for organizations. Table 10.2 presents advantages and disadvantages for individuals.

The eight organizational advantages cited in the literature focus on cost reduction and increases in productivity. Associated with cost are decreased turnover and absenteeism. Managerial advantages include increased managerial skill, improved employee morale, and increased flexibility regarding pay and benefits. The disadvantages proposed mirror the advantages. These include loss of control over employees (presumably

TABLE   10.1   Advantages/disadvantages   of   telecommuting   for
organizations

	Authors
Advantages to organization	
Reduced office costs	Atkinson, 1985; Walkin, 1985; Kelly, 1985; Guiley, 1985
Increased productivity	Rifkin, 1983; Gluckin, 1985; Walkin, 1985; Olson, 1982a; Pollack, 1981, Sample, 1981; Atkinson, 1985; Kelly, 1985; Regenye, 1985; Guiley, 1985
Improved employee morale	Olson, 1982a; DeSanctis, 1983; Guiley, 1985
Decreased turnover	Olson, 1982a; Guiley, 1985; Gordon, 1984
Decreased absenteeism	Olson, 1982a,b; Kelly, 1985; DeSanctis, 1983; Konrady, 1985
Improved managerial skills	Zientara, 1983
Increased organizational flexibility re: pay and benefits	Wang Laboratories, 1984; Kelly, 1985
Disadvantages to organization	
Decreased quantity and quality of management feedback re: employee performance	Shamir and Salomon, 1985
Loss of organizational culture	Shamir and Salomon, 1985
Problems with employee self-motivation and loyalty	Nilles, 1982; Farwell, 1984; Walkin, 1985
Legal issues	Sample, 1981
Loss of control over employees	Olson, 1982b; Whalen, 1984
Loss of congruency between personal and organizational goals	Nilles, 1982

TABLE   10.2   Advantages/disadvantages   of   telecommuting   for
individuals

	Authors
Advantages to individual	
Decreased travel stress	Olson, 1982b; Foegan, 1985
Reduced transportation costs	Olson, 1982b; Antonoff, 1985; Shirley, 1985
Jobs suitable for homebound workers	Olson, 1982B; Magee, 1985; Dattilo, 1985; Gordon, 1984
Increased job satisfaction	Guiley, 1985; Wolfgram, 1984
Increased job flexibility	Magee, 1985; Gordon, 1984
Improved time management	Olson, 1982a
Increased flexibility of home life	Kelly, 1985; Antonoff, 1985
Disadvantages to individual	
Isolation	Whalen, 1984; Wolfgram, 1984; Wiegner and Paris, 1983; Nilles, 1982
Loss of social and work-related interactions	Olson, 1982b; Walkin, 1985
Inhibited job development	Olson, 1982b; Walkin, 1985
Decreased career potential and promotability	Olson, 1982a,b; Lallande, 1984
Increased stress in the home	Shamir and Salomon, 1985; Nilles, 1982; Becker, 1981; Olson, 1983
Decreased job satisfaction	Dattilo, 1985
Lack of worker self-discipline	Dattilo, 1985; Farwell,1984
Worker exploitation	Dattilo, 1985; Antonoff, 1985; Lallande, 1984

by managers who have not increased their skills), loss of organizational culture, problems with motivating employees and divergence of personal from organizational goals. It is interesting to note in passing that most of the cited advantages are quantitative, while the majority of the cited disadvantages are qualitative.

Table 10.2 presents the advantages and disadvantages to the individual. A number of the advantages, such as reduced transportation costs and increased flexibility in home life, are particular to the individual. Others, such as decreased travel stress and increased job satisfaction, are of potential benefit both to the individual and the organization. The disadvantages for the individual reflect some of the controversy in the literature. These include decreased job satisfaction and increased stress in the home. Isolation and loss of social and work-related interactions seem particular to telecommuting.

The studies cited here do not form a seamless whole. There are a number of conflicting results and conclusions. For example, in one study, Olson (1985) concluded that employees need to 'feel a sense of belonging that is gained by walking through office doors.' While this was true for Olson's sample, very likely it is not true for all individuals. For example, employees in jobs which have little social contact may feel alienated at work.

Another conclusion drawn from the literature is that limitations to career potential and promotion are a major disadvantage for telecommuters (DeSanctis, 1983; Nilles, 1982; Olson, 1983). For many workers, however, the issue of a career and promotion is unimportant or irrelevant. Many simply do not have the skills, desire or opportunity for advancement.

Job satisfaction is another contentious issue. Some authors such as Guiley (1985) and Wolfgram (1984) argue that the increased personal flexibility and control associated with telecommuting will lead to higher job satisfaction. Other authors such as Dattilo (1985) and Shamir and Salomon (1985) argue that social isolation may make the home an undesirable workplace and result in decreased satisfaction. They conclude that telecommuting will not, in general, improve the quality of working life for employees.

Ramsower (1985) recognized the problems of making general conclusions. In a study of 16 telecommuters in five organizations he concluded that full-time telecommuting produces many negative organizational and behavioral effects. He further notes, however, that these effects can be overcome by both a strong desire on the part of the telecommuter to work in the home, and a willingness on the part of the organization to limit the employee's responsibilities.

One reason for the variety of viewpoints is the lack of empirical data. In the 56 articles cited here, only a few authors, such as Duxbury, Olson and Ramsower, had empirical data. Most of the articles are either anecdotal, are based on reviews of other authors, or are speculative. The main reason for this weakness is that full-time remote work or telecommuting is fairly rare. For example, in her 1985 survey in Datamation, Olson found that only 6% of respondents reported they were telecommuting full-time. In a similar vein, Duxbury, Higgins and Irving (1987) reported that only 3.5% (six) of 177 organizations had programs which allowed workers to stay at home for at least part of the day. Consequently, much of the literature is conceptual and is based on differing viewpoints. Given this situation it is not surprising that conflicting claims are made for telecommuting.

A further source of confusion arises from the differing nature of the articles. For example, the Duxbury study collected the attitudes of managers and professionals to telecommuting. The attitudes and opinions of this specalized group provide some measure of how attractive telecommuting is *a priori*, but do not address the reactions of experienced telecommuters. Other articles such as Batt (1982), Cheatham (1983) and Foegen (1985) present an overview of telecommuting for a general audience. Consequently they summarize the results and opinions of other writers but add little that is new to the discussion.

In a detailed study of telecommuting in one organization, Irving, Higgins and Macdonald (1986) found that the success of the program was a function of its attractiveness both to workers and managers. The factors that were most important to the organization were productivity and overhead (i.e. cost). The factors that were most important to the employees were quality of family life, quality of work life and commuting costs. It is important to note that the employees were all clerical workers who 'had jobs not careers.'

GUIDELINES FOR MANAGERS

Given the diversity of the field it is difficult to develop detailed guidelines for managers who must deal with flexible work in its many forms. The guidelines which follow are necessarily broad, and are focused on the key managerial issues listed in Table 10.3. These four issues are derived from the literature summarized in Tables 10.1 and 10.2. Individual managers must interpret them in the context of their own organizations.

TABLE 10.3   Key management issues

(1) What level of flexible work arrangements, if any, is appropriate for our organization?
(2) Which employees should be involved in which arrangement?
(3) What managerial structures should be in place to effectively manage our flexible work arrangements?
(4) How should our flexible work arrangements be evaluate

Issue 1: What Level of Flexible Work Arrangements, If Any, is Appropriate for our Organization?

Based on our earlier discussion it seems that full-time telecommuting should be considered only for clerical workers, some computer professionals and contract workers (e.g. consultants). The key to deciding feasibility is the closeness of coupling between a particular job or task and other organizational tasks or jobs, the degree of personal contact required by the job, the level of organizational security associated with the job or tasks, and the need for direct supervision.

One illustration results from the comparison of the job tasks assigned to a data entry clerk and a secretary. The key is to note how often one interacts with these employees during the day and how often they must refer to resources permanently located on the company's premises. Data entry clerks frequently receive reports or data to be entered in large groups. Further, they tend to work through this process with little interruption or questioning. The exception is unclear data entered from internal resources. In this case the entry clerk must frequently clarify items to be entered. If the former is true, then it does not make any difference if the employee is on the company's premises all day or not. He or she would be an ideal candidate for telecommuting.

By contrast consider an executive secretary. This employee frequently interacts with his or her boss, receiving last-minute requests and updating the boss on incoming phone messages and subordinates' requests for meetings, etc. This person would not be a likely candidate for telecommuting because he or she has high connectivity.

An accounting clerk illustrates another example. Although an accounting clerk may spend the bulk of the day processing accounting entries, he or she frequently has to refer to hard-copy files on the company's premises for backup documentation or clarification. In the case of even a medium-sized company these files can take up many filing cabinets. It is not feasible to duplicate these records in the accountant's home or on other premises. This person is also not an ideal candidate for telecommuting.

For those jobs where full-time telecommuting is feasible, one should assess how desirable it is. The desirability of full-time telecommuting has organizational and individual components. The organizational aspect centers around cost and productivity.

Other related issues include security of corporate data and equipment, accident insurance for work-at-home employees and procedures/policies for recovering company property if a work-at-home employee's contract is terminated. The individual aspects center around an employee's suitability for telecommuting and his or her willingness to do it.

Part-time telecommuting raises most of the same issues that full-time does. Extended day and extended week telecommuting require less investigation. The issues mainly center around theft insurance for the equipment, and policies to control use of corporate data.

Issue 2: Which Employees Should be Involved in Which Arrangement?

Identifying employees who may participate in flexible work arrangements is as important as defining the appropriate arrangements. In the previous section we identified likely candidates for full-time telecommuting. Clearly executives and professionals are likely candidates for extended day and extended week telecommuting. In addition some employees, such as salesmen, may already work remotely. Providing increased support in terms of cellular phones, portable fax machines and laptop (or notebook) computers is likely a wise investment.

Part-time remote work is often used as a method of retaining the skills of a valuable employee who otherwise would be lost to the organization either temporarily or permanently. Frequently, parents who must return early from maternity or paternity leave, or employees who become disabled, are involved in this type of arrangement (or in short-term, full-time telecommuting). In major cities where many employees face a long and expensive commute to work, full- or part-time telecommuting may be perceived as a job benefit. Even if the organization is in an undesirable location it can attract employees who would otherwise not be interested in working for the organization.

The other key consideration in identifying likely employees is the selection of those who will be successful telecommuters. Again, the selection of people for full-time telecommuting is more critical than for other arrangements. Based on the research to date it seems clear that the most likely candidates are those who do not have strong social needs, who have some experience with the organization and who have incentives to be at home.

For extended day and extended week arrangements the corporate donation of a computer and modem may be seen as a perk; particularly for those who currently commute to the office at weekends. Even where home computers are not used extensively for office work, the learning that takes place may pay off in the office. Many employees who have a computer at home use it to upgrade their computer skills on their own time. This learning is then translated into greater productivity back at the office.

Issue 3: What Managerial Structures Should be in Place to Effectively Control our Flexible Work Arrangements?

In order to manage the varieties of flexible work arrangements effectively, managers must have a clear vision of their purpose and goals. The first step is to assess how performance is currently measured in the organization. Given the concerns expressed by both managers and employees regarding monitoring (see chapter 9), it is clear that task monitoring must be outcome-based not process-based. In other words, employees must be assessed on the final outcome rather than on the process. In addition, pay, advancement and support structures must be clearly thought out; for example, a hot line for telecommuters is essential to provide quick support.

Other issues include reimbursement of telecommuters for time lost due to technical difficulties, a schedule of visits to the office for full-time tele-commuters, and training managers to deal with remote employees. Finally, some assessment of the technical infrastructure must be attempted. In some subdivisions, antiquated telephone switching equipment may preclude effective use of a modem. In some areas, having as few as 25–30 telecommuters can tie up a PBX, which is typically designed for calls of three minutes or less.

Issue 4: How Should our Flexible Work Arrangements be Evaluated?

We suggest that flexible work arrangements be evaluated in terms of the payoff to the organization and the satisfaction of employees who participate in these programs. As is the case with computerized monitoring, end-user computing and other issues, flexible work must be re-evaluated periodically and adjusted where necessary. Clearly flexible work schedules are not for everyone all the time. New employees and those with high social needs, or those who jobs require a high degree of interaction, should be at the office. For employees who have isolated jobs, those who must be at home for considerable periods of time, or those who find commuting disagreeable, flexible job arrangements may be appropriate. New parents, employees near retirement, and those who are physically disabled may find some level of telecommuting appropriate. Organizations faced with high costs of space may see flexible work in its variety of forms as an attractive alternative.

BIBLIOGRAPHY

Alderfer, C. P. (1972) *Existence, Relatedness and Growth: Human Needs in Organizational Settings*, Free Press, New York.

Antonoff, M. (1985) 'The push for telecommuting', *Personal Computing*, July, pp. 82–92.

Atkinson, W. (1985) 'Home/work', *Personnel Journal*, **64**(11), 105–109.

Batt, R. (1982) 'Fairchild giving "Telecommuting" a try in Phoenix', *Computerworld*, **16**(17), 71–72.

Becker, F. (1981) 'Mixed blessings: the office at home', *AFIPS Office Automation Conference*, Texas, 23–25 March, pp. 199–203.

Benson, David H. (1988) 'A field study of end user computing: findings and issues', in R. R. Nelson (ed.), *End-user Computing—Concepts, Issues and Applications*, Wiley, Toronto.

Business Week (1981) 'The potential for telecommuting', *Business Week*, **94**(98), 94.

Chabrow, E. (1985) 'Telecommuting: managing the remote workplace', *Information Week*, April, p. 28.

Chamot, D. and Zalusky, J. L. (1985) 'Use and misuse of office workstations at home', *Office Workstations in the Home*, National Academy Press, Washington, DC.

Cheatham, C. (1983) 'New workplace techniques: can they benefit your firm?', *Woman CPA*, **45**(2), 3–5.

Cheney, P. H., Mann, R. I. and Amorosso, D. L. (1988) 'Organizational factors affecting the success of end-user computing', in R. R. Nelson (ed.), *End-user Computing—Concepts, Issues and Applications*, Wiley, Toronto.

Connolly, S. (1988) 'Homeworking through new technology: opportunities and opposition—Part One', *Industrial Management Data Systems (IMDS)*, September–October, pp. 3–8.

Connolly, S. (1988) 'Homeworking through new technology: opportunities and opposition—Part Two', *Industrial Management Data Systems (IMDS)*, November–December, pp. 7–12.

Cowan, W. (1983) 'How administrators view the work-at-home trend', *Office Administration and Automation*, **44**(11), 28–105.

Dattilo, L. (1985) 'Back to homework', *Datamation*, **31**(3), 160–162.

DeSanctis, G. (1983) 'A telecommuting primer', *Datamation*, **29**(10), 214–220.

Downing-Faircloth, M. (1982) 'Would working at home be wise', *Personal Computing*, May, p. 42.

Drucker, P. (1973) 'Evolution of the knowledge worker', in F. Best (ed.), *The Future of Work*. Prentice-Hall, Englewood, NJ.

Dullea, G. (1983) 'New marital stress: the computer complex', *New York Times*, 10 January, p. A17.

Duxbury, L. E., Higgins, C. A. and Irving, R. H. (1987) 'Attitudes of managers and employees to telecommuting', *INFOR*, **25**(3), 273–285.

Economist (1987) 'Staying away in droves', *Economist*, 4 April, **303**(7492), p. 88.

Eder, P. (1983) 'Telecommuters: the stay-at-home work force of the future', *Futurist*, June, pp. 30–35.

Eisen, M. (1984) 'Business computing in the home-how you can make it happen', *Computer Dealer*, March, pp. 64–65.

Engstroem, M. G., Paavonen, H. and Sahlberg B. (1986) *Tomorrow's Work in Today's Society*, Swedish Council for Building Research, Stockholm, Sweden.

Farwell, B. (1984) 'Telecommuting and organizational change', *Computer Networks*, **8**(3), 169–173.

Flemming, D. (1988) 'A design for telecommuting', *Personal Computing*, October, pp. 148–150.

Foegen, J. (1985) 'The new cottage industries create new issues in benefits', *Personnel Journal*, **64**(2), 28–30.

Geisler, G. (1985) 'Blue Cross/Blue Shield of South Carolina: program for clerical workers', *Workstations in the Home*, National Academy of Sciences, Washington, DC, pp. 16–23.

Gluckin, N. (1985) 'The office is where the workers are', *Telecommunication Products and Technology*, **3**(6), 56–60.

Gordon, G. (1984) 'The office away from the office', *Computerworld*, **18**(38), 1–8.

Gregory, J. (1985) 'Clerical workers and new office technologies', *Office Workstations in the Home*, National Academy of Sciences, Washington, DC, pp. 112–124.

Grieves, R. (1984) 'Telecommuting from a flexiplace', *Time Magazine*, **57** (January).

Guiley, R. (1985) 'When your employees work from home', *Working Woman*, **10**(3), 27–30.

Heilmann, W. (1988) 'Organizational development of teleprogramming', in W. B. Korte, S. Robinson and W. Steinle (eds), *Telework: Present Situation and Future Development of a New Form of Work Organization*, Elsevier, Amsterdam.

Higgins, C. A., Irving, R. H. and Macdonald, K. (1988) 'Two Perspectives on Telecommuting', August, paper presented at Canadian Psychological Conference.

Hollis, R. (1984) 'Computer won't end commuting', *Toronto Daily Star*, 6 May, p. G4.

Huws, U. (1984) 'New technology homeworkers', *Employment Gazette*, January, pp. 13–17.

Information Systems (1984) 'The pluses of working at home', *Information Systems*, March, pp. 26–37.

Irving, R. H., Higgins, C. A. and Macdonald, K. (1986) 'Telecommuting: preliminary results', research report presented to The Canadian Workplace Automation Research Center, Department of Communications, Montreal, 12 June.

Jacobs, S. (1984) 'Working at home electronically', *New England Business*, **6**(9), 27–30.

Kelly, M. (1985) 'The next workplace revolution: telecommuting', *Supervisory Management*, **30**(10), 2–7.

Kelly, M. (1988) 'The work-at-home future', *Futurist*, November–December, pp. 28–32.

Konrady, E. (1985) 'How to manage the electronic cottage', *ICP Data Processing Management*, **10**(2), 15–15.

Korte, W. B. (1988) 'Telework—potential, inceptions, operations and likely future', *Telework: Present Situation and Future Development of a New Form of Work Organization*, Elsevier, Amsterdam.

Kraemer, K. (1982) 'Teleworking substitution for intracity travel', *Telecommunications Policy*, March, pp. 53–59.

Kroeber, D. and Watson, H. (1984) *Computer Based Information Systems*, Macmillan, New York.

Lallande, A. (1984) 'Probing the telecommuting debate', *Business Computer Systems*, **3**(4), 102–113.

Lipsig-Mumme, C. (1983) 'The renaissance of homeworking in developed economies', *Relations Industrielles*, **38**(3), 545–565.

Long, R. J. (1987) *New Office Information Technology: Human and Managerial Implications*, Croom Helm, London and New York.

Lopez, D. and Gray, P. (1977) 'The substitution of communication for transportation—a case study', *Management Science*, **23**(11), 1149–1160.

Magee, J. (1985) 'SMR forum: what information technology has in store for managers', *Sloan Management Review*, **26**(2), 45–49.

Mattera, P. (1982) 'Home computer sweatshops', *Nation*, 2 April, pp. 390–392.

McGee, L. F. (1988) 'Setting up work at home', *Personnel Administrator*, December, pp. 58–62.

Metzger, R. and Von Glinow, M. A. (1988) 'Off-site workers: at home and abroad', *California Management Review*, Spring, pp. 101–111.

National Academy of Sciences (1985a) 'Discussions: labor issues', *Workstations in the Home*, pp. 85–94.

National Academy of Sciences (1985b) 'Discussions: lessons learned', *Workstations in the Home*, pp. 95–104.

Newsweek (1989) 'Escape from the office', *Newsweek*, 24 April.

Nilles, J. (1982) 'Teleworking: working closer to home', *Technology Review*, **85**(3), 56–62.

Nilles, J., Carlson, F., Gray, P. and Hannemann, G. (1976) 'Telecommuting—an alternative to urban transportation congestion', *IEEE Transactions of Systems, Man and Cybernetics*, **SMC-6**(2), 77–84.

Olson, M. (1982a) 'New information technology and organizational culture', *Management Information Systems Quarterly*, Special Issue, pp. 71–92.

Olson, M. (1982b) 'Statement of concern', *Office Technology and People*, **1**(1), 37–40.

Olson, M. (1983) 'Remote office work: changing work in space and time', *Communications of the ACM*, **26**(3), 182–187.

Olson, M. (1985) 'Do you telecommute', *Datamation*, **31**(20), 129–132.

Olson, M. (1988) 'Organizational development of teleprogramming', *Telework: Present Situation and Future Development of a New Form of Work Organization*, Elsevier, Amsterdam.

Olson, M. and Lucas, H. C. Jr. (1982) 'The impact of office automation on the organization: some implications for research and practice', *Communications of the ACM*, **25**(11), 838–847.

Panko, R. R. (1987) 'Directions and issues in end-user computing', *INFOR*, **25**(3), 181–197.

Panko, R. R. (1988) *End User Computing*, Wiley, Toronto.

Pollack, A. (1981) 'Rising trend of computer age: employees who work at home', *New York Times*, 12 March, p. A1.

Post, D. (1982) 'Telecommuting: toward an extended office', *Interface Age*, **7**(10), 60–67.

Pratt, J. H. (1984) 'Home telecommuting: a study of its pioneers', *Technological Forecasting and Social Change*, **25**(1), 1–14.

Ramsower, R. M. (1985) *Telecommuting: The Organizational and Behavioral Effects of Working at Home*, UMI Research Press, Ann Arbor, MI.

Raney, J. G. (1985) 'American Express company: project homebound', *Workstations in the Home*, National Academy of Sciences, Washington, DC, pp. 8–15.

Regenye, S. (1985) 'Telecommuting', *Journal of Information Management*, **6**(2), 15–23.

Renfo, W. (1982) 'Second thoughts on moving the office home', *Futurist*, June, pp. 45–48.

Rifkin, G. (1983) 'Working remotely: where will your office be?', *Computerworld*, **17**(24A), 67–74.

Ronen, R. (1984) *Alternative Work Schedules: Selecting, Implementing and Evaluating*, Dow Jones Irwin, Holmwood, IL.

Rossman, M. (1983) 'Of marriage in the computer age', *Creative Computing*, August, pp. 132–137.

Sales and Marketing Management (1988) 'Work at home potential booms', *Sales and Marketing Management*, January.

Salomon, I. and Salomon, M. (1984) 'Telecommuting: the employee's perspective', *Technological Forecasting and Social Change*, **25**(1), 15–28.

Sample, R. (1981) 'Coping with the "work-at-home" trend', *Administrative Management*, **42**(8), 24–27.

Schiff, F. W. (1983) 'Flexiplace Pros and Cons', *Futurist*, **17**(3), 32–33.

Shamir, B. and Salomon, I. (1985) 'Work-at-home and the quality of working life', *Academy of Management Review*, **10**(3), 455–464.

Shirley, S. (1985) 'A company without offices', *Harvard Business Review*, **64**(1), 127–135. Telework: Impact on Living and Working Conditions (1984) European Foundation for the Improvement of Living and Working Conditions: Dublin. Published Report.

Toffler, A. (1981) *The Third Wave*, Bantam, New York.

Tzivanis Benham, B. (1988) 'Telecommuting: there's no place like home', *Best's Review*, May, pp. 33–38.

Vitalari, N., Venkatesh, A. and Gronhaug K. (1985) 'Computing in the home: shifts in the time allocation', *Communications of the ACM*, **28**(5), 512–522.

Walkin, E. (1985) 'Telecommuting: today's home work', *Today's Office*, **20**(7), 23–28.

Wang Laboratories Inc. (1984) *Telecommuting*, Advanced Systems Laboratory, Boston, Mass.

Whalen, B. (1984) 'Telecommuting: not a curious novelty but a trend with marketing applications', *Marketing News*, **18**(1), 14–16.

Wiegner, K. and Paris, E. (1983) 'A job with a view', *Forbes*, 12 September, pp. 143–150.

Wolfgram, T. (1984) 'Working at home: the growth of cottage industry', *Futurist*, **18**(3), 31–34.

Zientara, M. (1983) 'Exec: telecommuting strengthens managers' skills by requiring more active supervision', *Computerworld*, **17**(3), 28–29.

Chapter 11
END-USER COMPUTING

The 1980s ushered in dramatic changes in computer technology. Hardware costs dropped dramatically and software developers delivered easy-to-use, powerful systems that met the needs of many businesses. Personal computers, fourth-generation languages, relational databases, expert systems, and networks became common business tools.

The advances in hardware and software were matched by rising levels of computer literacy among managers and professionals. Computers were no longer viewed as black boxes that needed magicians to operate them. In this environment it is not surprising that untrained users began to develop applications programs, a task that was once the domain of the data processing department. The 1980s witnessed the birth of end-user computing; the 1990s will see its maturity. Rockart and Flannery (1983) report that the organizations they studied experienced a growth rate of 50–90% in end-user computing, while their data processing systems experienced a 5–15% growth. Even when we allow that EUC is relatively new, and DP is relatively mature, this difference in growth is remarkable.

End-user computing has a variety of definitions. We define EUC as the process by which users (individual or group) develop their own applications(s) instead of relying on a central information systems group. At one end of the spectrum the end-user simply enters command-level instructions (such as the menu system in Lotus 1–2–3). The other end of the spectrum involves a complete systems development similar to the process used by information systems personnel.

End-user computing emerged as a managerial issue in the early 1980s with the proliferation of personal computers and the parallel growth of sophisticated fourth-generation languages. (A fourth-generation language (4GL) is a computer language that has the following characteristics: it has an English-like grammar, it is non-procedural, it requires much less learning time and programming time than preceding generations.) These hardware and software systems allowed relatively inexperienced users to write very sophisticated programs. More importantly, they gave users control over their data processing needs.

Traditionally, an organization's data processing needs were met by an information systems group. This group developed applications using formal systems analysis techniques and quality control standards with technical experts in control of the complete data processing cycle. While this approach to developing applications programs had technical advantages, it left too much control in the hands of people without in-depth business knowledge. Many systems developed by technical experts turned out to be unsatisfactory. Furthermore, the demand for applications development projects increased to the point where in the late 1970s most IS groups were working with a two-year backlog. Clearly something had to be done.

The development of user-friendly, powerful software reduced the requirements for technical expertise to a level where many managers were able to develop their own software systems. End-users began to take over much of their own programming and applications development work.

The development of easy-to-use end-user application tools combined with the (perceived) difficulty in working with IS professionals led to rapid growth in EUC. In a survey of key issues facing IS managers, Dickson *et al.* (1984) found that the management of EUC was the second most important issue. More recent surveys, however, rate EUC much lower in importance, showing the increasing maturity of the field.

Benjamin (1982), the International Data Corporation (1984) and others estimate that during the early 1990s EUC will be bigger than data processing. Clearly this is an issue that will assume increasing importance to managers over the next decade. It is in this context that we examine its advantages and disadvantages, and discuss how to manage EUC effectively.

ADVANTAGES OF END-USER COMPUTING

End-user computing provides many advantages to the organization; these include:

(1) control in hands of ultimate users;
(2) reduction of the IS backlog and application maintenance load;
(3) an increase in efficiency and effectiveness;
(4) improved communication and appreciation between IS department and user groups.

The primary advantage of EUC is that it transfers control for applications development to end-users who are most familiar with their own business needs. It allows the end-user to control which data are to be used, which analysis is to be done and how the output should be tailored for presentation. With control of the development process the end-user can get more timely analysis of the business problem without the frustration of continual interactions with another department (i.e. IS) outside his/her control.

End-user computing can also reduce the large backlog of IS applications that are typical of many data processing shops. Estimates in the early 1980s put the backlog at two years. Today, with EUC applications, backlogs have shrunk substantially. In many organizations there has been a down-sizing in IS staff as more and more applications are off-loaded to the final users.

End-user computing is believed to increase productivity. During the recession of the early 1980s many organizations turned to technology as a means to decrease costs and increase productivity. Their investments are beginning to pay off. Current estimates are that EUC will result in a productivity gain of 10–50%. This will be discussed in the next section.

Several studies (e.g. Gerrity and Rockart, 1986) have shown that EUC has a positive effect on communications between IS staff and end-user groups. This improved communication results in less friction between the groups and in the development of better decisions.

DISADVANTAGES OF EUC

Despite the advantages of EUC, management must consider several potential dangers. These include:

(1) uncertain quality of the applications program;
(2) uncontrolled proliferation of hardware/software;
(3) duplication of system development;
(4) unanticipated training costs and time;
(5) threats to data security;
(6) hidden productivity losses.

A major concern associated with EUC is the quality of the applications program. End-users are not trained, as are IS specialists, in the development, testing and documentation of systems. Thus, when an end-user develops his/her own applications program, it is often not thoroughly tested or properly checked for errors. When one sees the output from an application developed by an end-user, there is no assurance that the data are accurate, that the manipulations of the data are correct or that the assumptions underlying the analysis are tenable. In fact, one study estimated that over 50% of applications developed by end-users had a flaw in the data or the analysis.

Quality control of user-developed applications is particularly important in situations where several people or departments may be using the application or its output. Furthermore, if the system is used repeatedly, quality control is important because not only can an error occur, but the error can be multiplied and reapplied over that length of time. If the output from the applications is going to support a critical decision, if the system requires a large database, or if the application involved acts as a modification to a corporate database, then the quality control of end-user applications is an absolute necessity.

Related to this issue is a problem associated with documentation and backup. As stated earlier, end-users are not trained IS personnel and do not know how to document their systems adequately. This is acceptable if individuals stay with the organization, but when they leave the organization may be left with a legacy of systems about which they know nothing.

One solution is to set the same documentation standards and requirements for users as for the systems staff. The catch is that documentation standards should never be so stringent that they are overly onerous to follow. By encouraging the user to create the systems and explaining the need for documentation and the benefits for them (e.g. so someone can run the system while they are on vacation or so that systems can help them solve overly technical problems), users can be made to feel part of the overall system development process. Users who are left on their own will feel isolated and insecure in their attempts to develop applications. They will not take ownership of any documentation requirements if they do not feel there is a benefit for themselves.

Underlying many problems with EUC is a halo effect associated with computer output. According to this well-known phenomenon, if people are generally impressed with an object (usually a person), they will tend to ignore the less desirable attributes of the object (DeSoto, 1961). Thus, if people are impressed with computers they will not question the output from these computers, however the output was derived.

While many managers reject the idea of a halo effect, our own research (Higgins, Huff and Lin, 1987) gives a strong indication of its presence. In this study we had managers analyze a business decision of a toy manufacturer. All managers examined exactly the same business decision with exactly the same information. In half the cases the managers had computer output to look at. The other managers had exactly the same output done on a typewriter.

Interestingly, the managers with the computer output gave a significantly higher score when asked about the success of the toy manufacturer, clearly indicating a halo effect associated with the computer output. Furthermore, the less experienced a person was with computers the higher the score they gave to the potential success of the toy manufacturer.

A second disadvantage of EUC is the potential proliferation of hardware and software. It is not uncommon for organizations to have many incompatible hardware and software systems. This increases support costs as IS has to learn many different vendor configurations in order to repair and integrate the various systems. Furthermore, there are increased costs of duplicating peripheral equipment or software add-ons if different vendors cannot work with common equipment. For example, some vendors require that their own printer is hooked up to their equipment. Different equipment may then mean purchasing more printers than are absolutely necessary for workload reasons, just to be able to print from all machines.

A third problem associated with EUC relates to training. Many organizations did not anticipate the large resources they are spending (through information centers and the like) or will continue to spend in the future. Furthermore, the highest costs of EUC are associated with novices attempting complex inquiries. Organizations must be diligent in order to prevent users from avoiding proper training and wasting inordinate amounts of time teaching themselves (usually improperly) how to use certain software packages.

A fourth problem associated with EUC concerns security. Employees with access to company databases can easily obtain copies of vital company data. Furthermore, as organizations make their computing capabilities accessible to employees who want to work from home in the evening, the threat of outsiders breaking into the computer increases dramatically. Home computers do not generally have the same extensive security limitations as those computers in the office, for example login procedures and restricted physical access. Also more copies of the data lead to a higher probability of some falling into the wrong hands. It is easy to take a copy of data and leave no trace of there having been an infringement of security; also it only requires a disk and a rudimentary knowledge of the computer operating system to duplicate the data. This is easier than removing original copies and searching out a photocopier (especially if the data are also retained on home computers).

A fifth problem associated with EUC is productivity. Earlier, we identified productivity as an advantage. However, a strong case can be made that EUC, if inappropriately applied, is counterproductive. The classic case occurs when an end-user becomes obsessed with the technology to the point that it dominates his/her job rather than being a supporting mechanism. The end-user over-analyzes the problem and goes about an inefficient search for a solution. All organizations are familiar with individuals who continually tinker with their applications, never seeming to bring it to production. This is clearly counterproductive.

MANAGING EUC: WHAT CAN MANAGERS DO?

Most organizations facilitate EUC through an information center (IC). The concept of an IC was developed in the early 1980s by IBM to meet the growing needs of end-users. Information centers provide many of the following services. They:

(1) assist in the development of applications;
(2) provide support and education;
(3) act as an information clearing-house;
(4) foster user self-sufficiency;
(5) set hardware and software standards;
(6) help end-users justify and evaluate projects;

(7) facilitate the integration of end-user applications with business goals;

(8) ensure data integrity;

(9) provide data management services.

In short, the IC should be capable of providing end-users with the full range of support they need to develop useful, reliable business applications. They are generally only as good as the people selected to staff them; yet many organizations overlook the importance of putting their very best people in these groups.

A good starting point for the organization is to develop an EUC strategy. Furthermore, the organization must take an active role in targeting critical end-user systems and applications. Above all the process must be driven by a comprehensive, end-user education program.

Recently, some organizations have begun to disassemble their ICs. One reason is that their duties degenerate into solely training users on currently approved and used packages. Training initially does take a lot of time. However, as training needs decline (as a saturation point is reached) ICs do not always take the initiative to begin research into new and more effective technological processes for their organization. They get too focused in maintaining expertise in the training area, and other areas of the company get used to seeing them in this light.

We feel this is a mistake. Information centers should evolve to meet the ever-growing needs of an organization's computing. They can act as a technology watchdog, always looking for new products that fill a business need. Furthermore, ICs can take a proactive role in EUC by assisting users in identifying strategic requirements and then developing systems to meet them. A study by Cougar (1986) identified characteristics of successful and unsuccessful ICs. Successful ICs used soft controls; these include limiting support to approved software and hardware and offering extra support to complying end-users. By way of contrast, Cougar found that firms with very strict controls did little better than those with no controls at all. The reason seems to be that organizations with tight controls had little ability to enforce their restrictions. It seems that a combination of controls with incentives creates an appropriate environment for management of EUC.

In addition to maintaining an IC, development of an EUC user group is recommended. One approach is to identify an individual in each department or major work group in the organization, and require that person to take a common training course in applications development on personal computers. These people are then responsible for data control and data integrity in their area; they are also expected to act as resource persons for other end-users in their department or workgroups.

Collectively these user representatives are members of the EUC users' advisory committee that works with the IC to develop policies and guidelines for management of EUC. All end-users are members of the user group and can make their ideas known to this group. Representatives from this group will then negotiate with the IC, and with management if necessary. The advantage of this approach is that it disperses knowledge and expertise out to the firing line where it is most required. A drawback of this approach is that it ties up the time of company personnel who could be doing other things. Consequently, it can be resource-intensive. Of course, when one considers the possible cost of a major mistake, the costs associated with using personnel in this manner may seem small.

The organization should produce a set of guidelines (in consultation with the IC and the EUC user group) for the types of applications that are appropriate for end-user development. A good start is to look at the extensiveness of the applications. Are end-users developing programs that will be limited to their own use? Do the programs contain sensitive data or sensitive output? Are the programs going to be shared among several people? In a 1983 study, Rockart and Flannery found that only 31% of EUC applications were used by a single person, 52% were used by several people in the same department and 17% were used by people outside the department. Furthermore they found that 9% were large enough to be classified as operational systems, and 50% of all projects were complex analysis projects. These findings led them to conclude that EUC had a third environment consisting of large end-user projects. Panko (1988, p. 28) provides a critical perspective on these findings. While agreeing with their general thrust, he argues that Rockart and Flannery likely overestimated the prevalence of third environment applications. Still, as he points out, the 'power user' is a recognized part of the end-user landscape today (for more detail on EUC see Panko, 1988).

At the managerial level it is not so important that one be able to identify all the issues pertaining to EUC. It is sufficient that individual managers are aware that they have to establish and enforce guidelines for EUC. Furthermore they must ensure that appropriate structures exist for the management of EUC. We have already discussed one such structure, the EUC user group. Once a manager has put the structure in place he or she has taken a large step towards the control of EUC in the organization.

To go beyond basic control of EUC one must monitor what is happening. It is a managerial responsibility to periodically (once or twice a year) raise issues of data loss, potential problems with data security or excessive duplication of effort. These investigations might be as simple as walking around and asking people what they are doing with their computer, and what applications they developed, to get a feel for how people are treating applications development and the attitudes towards it. On the other hand, it could be as sophisticated as hiring an outside consultant to do a full-scale evaluation of EUC.

A summary of the management issues discussed above is outlined in Table 11.1. End-user computing is a phenomenon that will continue to grow.

The problems associated with EUC will grow in parallel with it. To realize the benefits of this phenomenon, managers must take steps to ensure that reliable applications are developed, and that the security and integrity of corporate databases are not jeopardized.

TABLE 11.1   Managerial issues in EUC

(1) Develop a strong information center staffed with good people.
(2) Provide a mandate for the IC to take a strong proactive approach to managing EUC that involves:
    (a) soft controls—incentives as well as rules,
    (b) focus on priority business needs,
    (c) development of good relations with the EUC users' group.
(3) Develop a EUC user group consisting of all end-users in the organization. The EUC user group will be managed by a steering committee composed of a representative of each key workgroup or department, representatives from the information center, and at least one middle-level manager. The steering committee is a working committee that is responsible for drafting policies for managing EUC.
(4) Inform all managers of their responsibilities concerning EUC.

BIBLIOGRAPHY

Alavi, M. (1985) 'Some thoughts on quality issues of end-user developed systems', *Proceedings of the Twenty-First Annual CPR/BDP Conference*, May, pp. 200–207.
Benjamin, R. I. (1982) 'Information technology in the 1990s: a long range planning scenario', *MIS Quarterly*, June, pp. 11–31.
Benjamin, R., Dickinson, C. and Rockart, J. (1985) 'Changing role of corporate information systems officer', *MIS Quarterly*, September, pp. 177–188.
Chen, R. (1985) 'The trained-user as a systems analyst', *Journal of Systems Management*, July, pp. 38–40.
Cougar, J. D. (1986) 'E. Pluribus Computum', *Harvard Business Review*, September–October, pp. 87–91.
Davis, J. (1985) 'A typology of management information systems users and its implications for user information satisfaction research, *Proceedings of the Twenty-First Annual CPR/BDP Conference*, May, pp. 152–164.
DeSoto, C. B. (1961) 'A predilection for single orders', *Journal of Abnormal and Social Psychology*, **60**, 16–23.
Dickson, G. W., Leitheiser, R. L., Wetherbe, J. C. and Nechis, M. (1984) 'Key information issues for the 1980's', *MIS Quarterly*, September, pp. 135–147.
Ditlea, S. (1985) 'Befriending the befuddles', *Datamation*, June, pp. 102–104.
Friedman A. and Greenbaum, J. (1984) 'Wanted: renaissance people', *Datamation*, **30**(14), p. 134.
Gerrity, T. P. and Rockart, J. F. (1986) 'End-user computing: are you a leader or a laggard?', *Sloan Management Review*, **27**(4), pp. 25–34.
Guimaraes, T. (1984) 'The evolution of the information centre', *Datamation*, July, pp. 127–130.
Head, R. (1985) 'Information resource centre: a new force in end-user computing', *Journal of Systems Management*, **36**(2), 24.

Higgins, C. A., Huff, S. L. and Lin, J. T. (1987) 'Computers and the halo effect', *Journal of Systems Management*, **38**(1), 21–23.

International Data Corporation (1984) 'The paper blizzard', *USA Today*, April, p. 1.

Kahn, B. and Garceau, L. (1984) 'Controlling the microcomputer environment', *Journal of Systems Management*, May, pp. 14–19.

Kalogeras, G. (1984) 'Data base—the technical heart of an information centre', *Journal of Systems Management*, November, pp. 36–38.

Karten, N. (1985) 'Surviving the PC challenge demands vigorous management skills', *Data Management*, September, pp. 14–18.

Khosrowpour, M. (1985) 'MIS Leadership in transition', *Journal of Systems Management*, November, pp. 18–21.

Kliem, R. L. (1985) 'In-house training for microcomputer users', *Administrative Management*, December, pp. 50–51.

Lehman, J. (1985) 'Personal computing vs. personal computers', *Proceedings of the Twenty-First Annual CPR/BDP Conference*, May, pp. 97–102.

Leitheiser, R. and Wetherbe, J. (1985) 'The successful information centre, what does it take? Systems', *Proceedings of the Twenty-First Annual CPR/BDP Conference*, May, pp. 56–65.

Mintzberg, H. (1990), 'Manager's job: folklore and fact', *Harvard Business Review*, March–April, pp. 163–176.

Panko, R. R. (1988) *End User Computing: Management, Applications, and Technology*, Wiley, New York.

Parker, R. G. (1984) 'Microcomputers—an issue of control', *Business Quarterly*, Fall, pp. 32–34.

Rockart, J. F. and Flannery, L. S. (1983) 'The management of end-user computing', *Communications of the ACM*, **26**(10), 776–784.

Vogel, D. (1985) 'Office automation: end user impact', *Proceedings of the Twenty-First Annual CPR/BDP Conference*, May, pp. 165–171.

Vogt, E. (1984) 'Managing the PC revolution', *Datamation*, 15 November, pp. 113–114.

Part IV
OIS AND THE FUTURE

This part contains one chapter (Chapter 12). Here we look to the future and try to highlight where OIS is going over the next 10 years. Our purpose is to integrate the material contained so far in the book, and to extend these ideas to identify management challenges in the future. Fast-paced changes in technology and organizational concepts mean predictions about future issues are subject to a wide range of variables. However, while the details may vary, useful images of the general trends can be formed.

Chapter 12: Management and OIS: Where Do We Go From Here?	What is the state of OIS today and what are the future management issues?

Chapter 12
MANAGEMENT AND OIS: WHERE DO WE GO FROM HERE?

In this chapter we summarize the ideas presented so far, discuss the state of OIS today, and look to the future of OIS and the management issues it will generate over the next 5–10 years. The reader can use this chapter to get a quick overview of the contents of the book and a brief summary of the current state of the art with respect to OIS. The larger value lies in the attempt to stimulate the reader to think creatively about OIS and potential changes that managers must deal with in the future.

WHERE WE HAVE BEEN: OIS CONCEPTS AND IDEAS

During the course of this book, OIS were discussed in terms of conceptual issues, operational issues and developing issues. At this point it is useful to review and summarize material covered earlier.

Chapters 1–5 presented conceptual issues related to OIS. The key concepts raised were:

(1) OIS changes the nature of the business itself;
(2) OIS has indirect use for competitive advantage in most firms;
(3) The use of OIS has direct effects on the organization and the societal context in which it operates;
(4) Integration of OIS systems is a complex and technically difficult task;
(5) Successful use of OIS requires first that management have a vision of the corporation and where it should go; that they communicate this vision effectively to employees; and that the employees buy into the vision.
(6) Computers in general, and OIS in particular, act to turn organizational constraints into variables. Opportunities are limited only by the creative ability of management to change the nature of the organization.

While these points summarize the key ideas, it is important to note that the theme of Part I is understanding the ideas behind OIS. In other words, to be able to deal with the issues raised by computers and OIS, a manager must have a firm grounding in concepts of organization, organizational structure, organizational communication and the issues generated by OIS technology.

In Part II the concepts raised in Part I were extended to three operational areas: planning for OIS, needs analysis for OIS, and implementing and evaluating OIS. The key themes raised in Part II were:

(1) A fully developed OIS requires careful long-term planning.
(2) To develop a well-designed plan a thorough and appropriate needs analysis must be conducted.
(3) To conduct an appropriate needs analysis the focus must be initially on workgroups, on the tasks that workgroups accomplish, and on the links that workgroups must maintain with other workgroups to accomplish these tasks.
(4) The data from the initial needs analysis provide input to the more detailed planning process, particularly when used with the feasibility triangle and the organizational scan as discussed in Chapter 3.
(5) Implementing and evaluating OIS is a long-term process that begins at the earliest planning stages and continues throughout the life cycle of the system. To be able to evaluate a system properly, one must have clear goals and directions and establish performance measures. Clearly, this is a responsibility of management early in the planning process. Furthermore, one must be able to plan for the implementation process well before the equipment arrives in the office.
(6) Training employees to use the system and ongoing monitoring of their use of the system is a long-term expensive activity that needs careful management attention. A number of training and education issues that must be addressed were discussed in Chapter 8.

Part II focused on management action that can be taken to control the complete process of OIS design, development, implementation and evaluation. Note that the emphasis was on management action, not technical action. The reason for this focus was that it is senior management's responsibility to control the conceptual architecture that underlies the OIS system. This, in effect, is a reflection of the organizational philosophy of the company and must incorporate management's vision of where the organization is to go in the future. Consequently, the aim of Part II was to put management solidly in charge of this process.

In Part III we addressed three developing issues that managers, who have or plan to acquire OIS systems, must deal with. These were computerized performance monitoring, flexible work and end-user computing. The focus of Part III differed from the earlier two parts. In Part III the aim was to summarize current research and to use this information as a basis to extract management issues and actions that would assist in dealing with these developing issues. Other issues, which are in perhaps a more formative stage of development, are the use of expert systems by senior managers, use of decision support systems throughout the organization, integration of plant and office systems into true organizational information systems, and the development of electronically based groupworks by which several employees share a common 'information space.' While each of these issues is important, and each will become more so in the future, it was our feeling that there is not sufficient managerial research available at this time to include them.

The key points made in Part III were as follows:

(1) Employee participation in an active and meaningful sense in the design of monitoring systems, flexible work programs, and in developing end-user policies is a key factor in the success of any management actions taken.
(2) The use of extensive consultation, active user groups and ongoing assessment of current problems with each of these three developing areas is the key to maintaining effective management.
(3) While dialogue is important, it is also important that managers have a clear idea of the strategic value of computerized monitoring, flexible work, and end-user computing to their firm, and that they be willing to modify their policies, procedures and vision as the organizational environment changes. This necessitates an ongoing program of active environmental scanning.

WHERE WE ARE NOW: OIS TODAY

Currently OIS are in a transition phase. As was mentioned in Chapters 2 and 3, the level of integration and the sophistication of systems varies widely. There are emerging standards but no fully compatible communications or graphics. Even text standards have not yet been fully developed. Consequently, those who are using OIS and those who wish to develop them must focus on creating the maximum flexibility in current systems, and in devising architecture that allows for dramatic changes in technology over the next few years. Clearly, some recent changes have altered how office information systems are viewed. Among these are the development of cellular phones and portable facsimiles, and the improvement of laptop computers. As the technology develops it is clear that these capabilities will become sophisticated and refined. We also have much more sophisticated printing technology than we had 10 years ago, and better software. Not only that, a new breed of professionals and middle managers, who are computer-literate, are coming into the workforce. They expect computer technology to be available to them and will be a strong force for the development of more sophisticated OIS in the future.

Currently, the picture is both discouraging and encouraging. It is discouraging from the point of view of someone who recognizes the potential capabilities in comparison to what we actually have. It is encouraging in the sense that the users are demanding and getting a movement toward common standards for text, graphics and data communications. We still lack standards in training and education to help people use OIS effectively.

Most managers today are in a situation where they recognize that large changes can and perhaps should be made; on the other hand, they recognize that they have things to accomplish today and cannot wait for the ultimate technology that may be available in the next five to ten years. Furthermore, developing fully integrated OIS is still long-term, expensive and risky. Some systems will fail; some will succeed. All levels of management are obliged to recognize that they are operating at least in a moderate-risk environment and to take every precaution that they can to reduce the risk they cannot eliminate.

WHERE ARE WE GOING: OIS IN THE FUTURE

There is an old Chinese saying: 'prediction is difficult, particularly where it concerns the future.' This is particularly true where it concerns computer technology and its use in organizations. However, despite the risk there are some common themes that are developing. In a recent paper, Huber and McDaniel (1986) consider the central impact of the information society, and advanced information technology on design of post-industrial organizations. They conclude that the increasing penetration of computer technology in society will eventually create an increased pace of organizational work, increased fragmentation and increased demand or pressure to control decision processes. They argue that the next major organizational paradigm to develop will be management of decision processes.

One can currently see this trend in organizations that are highly computerized. The rate at which information is generated is increased, the rate at which decisions must be made is increased and the number of people who must and who can be consulted regarding any particular decision is increased. Consequently, developing effective procedures to manage and control decision processes is, Huber and McDaniel argue, the key to effective organizational design in management in the post-industrial era.

We extend Huber and McDaniel's ideas to some degree. Clearly, computer technology removes constraints. However, removing constraints without providing new directions for people to follow will result in electronic versions of existing systems being developed. As mentioned in Chapter 4, the key to effective use of computer technology is not just changing what people do, but in changing what people think about what they do. Consequently, organizations who are going to use advanced OIS effectively will, over the next five years, develop mechanisms for the development and adoption of new organizational visions. This means that they must be active scanners of their internal and external environments; must be creative synthesizers of new information; and must be able practitioners in the art of developing consensus.

Many of these skills exist today. However, in most organizations they are not given the prominence that they will acquire over the next few years.

Our view of OIS in the future is one in which managers will focus more on organizational processes and organizational outcomes; where discussion, consultation and creativity will of necessity be given more prominence than they are today; and where emphasis on bureaucratic thinking, following rules and accepting constraints, will be considerably reduced. To achieve this vision, managers must start today to take control of OIS technology and ensure that it is serving organizational goals.

BIBLIOGRAPHY

Huber, G. P. and McDaniel, R. R. (1986) 'The decision making paradigm of organizational design', *Management Science*, **32**(5), 572–589.

Appendix
AN HISTORICAL PERSPECTIVE ON PERFORMANCE MONITORING

In this section the development of organizational thinking on performance monitoring and control systems is traced, and presented according to the major divisions of managerial theory.

THE CLASSICAL SCHOOL

The classical or functional school of management attempted to discover basic principles of management akin to those being discovered in the physical sciences. The classical school of thought can be traced back to Charles Babbage who, in 1835, wrote a book describing the principles of the division of labor. His mechanistic approach foreshadowed the work of later writers such as Henri Fayol and Frederick Taylor. Incidentally, Babbage is also regarded by many as the Father of Computers, due to his work on a 'calculating engine.'

Henri Fayol, a French manager, published a book on general principles of management. His work, and the parallel work of Frederick Taylor on 'scientific management', continued the mechanistic paradigm of this industrial revolution. Workers were essentially machines to be controlled and directed. It was assumed that they had little ability to manage their own work and that, if properly trained, they would perform as directed.

During roughly the same period, Max Weber was developing his theory of bureaucracy in Germany. This macro-theory of organizational design is compatible with the works of Fayol and Taylor. What Weber accomplished was the design of a mechanism (bureaucracy) that allowed the extension of the concept of division of labor and scientific management to large organizations.

While Weber was founding the basis for administrative studies, Taylor's ideas and methods formed the basis of industrial engineering. Many of the initial work study methods are still used today. Robert Nolan (1983, pp. 111–297) describes the modern extensions of these theories (work measurement). A slightly different approach is taken by the human resource management school. This is described by Lehrer (1983, pp. 297–315).

The performance monitoring and control systems developed from these theories (i.e. work measurement/work standards, etc.) tend to be both rationalistic and mechanistic. They have had the most success when applied to highly structured, low-level jobs.

THE HUMAN RELATIONS SCHOOL

The human relations school developed as a consequence of the work of Elton Mayo and his colleagues at the Western Electric Hawthorne Plant. This school stresses the importance of the effect of informal and social relationships on productivity. In general, the school was concerned with relations between productivity and morale, leadership style and the nature of coordination between workgroups.

The emphasis on interpersonal relations led to research on group dynamics, leadership style and job design. One underlying assumption was that job satisfaction was directly related to productivity. However, research shows that this assumption is not valid, and that the relationship between job satisfaction and performance is indirect and complex (Howell, 1982). The performance monitoring and control system, based on a human relations approach, stresses interpersonal relations, individual growth, control of the workers over their environment, and job enrichment.

THE SOCIOTECHNICAL SCHOOL

This approach focuses on the design of the work situation based on the principle that the technical system must be compatible with the social system. Prominent in this school is the work of Trist and Bamforth (1951), Rice (1958), Klein (1976) and Hackman and Oldham (1980). Recently Kling and Scacchi (1982) have extended the sociotechnical approach to include organizational politics, as well as the more common task-oriented aspects of job design.

This school assumes that job performance is a complex function of: the design of the task, the nature of the individuals and the social matrix in which they perform the task, and the organizational control systems operating in that environment. Performance monitoring and control systems designed from this perspective should consider task, social, organizational and power aspects of any implementation (e.g. Markus, 1983).

CONCEPT OF CONTROL

Generally, there is agreement that established control standards serve as a basis for the comparison and evaluation of actual performance (Brown, 1968). The literature, however, presents many possible definitions for control and ways of classifying control systems. For example, control can mean:

(1) to direct, to influence or to determine the behavior of someone else (rooted in study of influence and power);
(2) the task of ensuring that planned activities are producing the desired results (Woodward, 1970; Eilon, 1962).

Control systems are classified in various ways, according to their purposes or the type of data collected. For example, Woodward suggests a scheme of classifying control systems based on two dimensions: personal–mechanical and unitary–fragmented. The personal–mechanical dimension refers to how control is exercised in an organization. A personal control system would be one in which the manager gives instructions to employees and monitors their work, while a mechanical system would include impersonal controls, such as cost or automatic control systems. The unitary–fragmented dimension reflects the degree to which the various managerial control systems of an organization are integrated.

Researchers appear to have viewed the concept of control from two perspectives:

(1) as a criticism of bureaucracy;
(2) as a study of individual psychology.

The criticism of bureaucracy emphasizes the dysfunctional consequences of control systems. For example, Merton (1940), in his model of such consequences, points to the resulting rigidity of behavior and increase in difficulty with clients. The individual psychological approach has its basis in various motivational theories (e.g. reinforcement—Skinner, 1953; needs—Maslow, 1954; expectancy—Lewin, 1938) and raises questions with respect to the individual's motivation on the job.

These approaches have stimulated a number of empirical studies on control systems. Some of their findings and implications are summarized. This review is, in part, based on Lawler's (1976) paper that presents a comprehensive review of the literature on control systems up to 1976.

IMPACTS OF CONTROL SYSTEMS ON BEHAVIOR

In connection with empirical studies, one should understand that in the opinion of some authors there are methodological problems associated with the existing research on impacts of control systems: 'There is a clear need for further theoretical explication and improved laboratory and field research aimed at enhancing both construct validity and substantive considerations' (Griffin, Welsh and Moorhead, 1981, p. 655).

Dysfunctional Effects

Most of the empirical studies on performance monitoring and control systems point to the dysfunctional effects of control systems. Although the purpose of a control system is generally to act as a means for ensuring that organizational goals are achieved, a control system can *prevent* the goals of the organization from being accomplished (Argyris, 1952; Blau, 1963; Brown, 1968; Zuboff, 1982).

Ridgway (1956, p. 243) points out that 'the existence of a measure of performance motivates individuals to *effort*, but the effort may be wasted ⋮ or may be detrimental to the organization's goal.' Ridgway also states that quantitative measures of performance, even if used purely for information purposes, are usually interpreted as the important aspects of the job. As a result these measures have important implications for the motivation of behavior. Such effects depend on the nature of the control system used, and the individuals in the organizations. Several types of dysfunctional behaviors can be caused by control systems. Some of these are listed below, followed by a discussion of each type:

(1) bureaucratic behavior,
(2) resistance,
(3) falsifying control system data.

These behavioral problems are believed to be greater when the information collected by the control system is used as a basis for a *reward system* (Kerr, 1982). This relationship is particularly strong in jobs where the level of trust is low, individual performance is difficult to measure, performance must be measured subjectively, or the system does not measure all the behavior that is needed for effective job performance. In order to understand how a system will impact on a person's behavior it is necessary to know how the information will be used and who will see it (Lawler and Rhode, 1976). One way to reduce the amount of resistance, to decrease the amount of bureaucratic behavior, and to increase the amount of valid data, is to tell employees that the data from the control system will not be used by the reward system (Lawler, 1976).

Hackman *et al.* report an example of selective elimination of controls in an insurance firm. When improvements in operator proficiency permitted them to work with fewer controls, the organization estimated that 'the reduction of controls had the same effect as adding seven operators—saving even beyond the effects of improved productivity and lowered absenteeism' (Hackman *et al.*, 1979, p. 100).

Bureaucratic Behavior

Studies have indicated that control systems may influence employees to perform in rigid ways that are dysfunctional as far as the goals of the organization are concerned. As an example, Blau's classic study of an employment agency indicated that employees were motivated to perform only those behaviors measured by the system (Blau, 1963). In this case the control system measured the number of interviews conducted, and used this information for evaluating individual workers. As a result the interviewers devoted their efforts to increasing the quantity of interviews, and the agency's goal (i.e. to place workers in jobs) was often ignored.

Resistance to Control Systems

Most authors explain the resistance to control systems in terms of their being perceived as a threat to the need satisfaction of employees (Argyris, 1952; Whisler, 1970). The amount of resistance is determined by the characteristics of both the control system and the individual workers. The resistance may not be based on *accurate* perceptions of what the impact of the system will be (i.e. as long as those perceptions *exist* there will be resistance).

The following list shows some of the reasons for resistance, as based on Lawler's review, with examples from other writers:

(1) *Control systems can automate expertise.* Systems that reduce the degree of expertise necessary to do a job, and that standardize work, appear to threaten the satisfaction of needs such as: security (the person may feel more expendable), status (what the person is respected for can become valueless), and autonomy (the system restricts a person's freedom to perform the job by requiring repetitive activities) (Argyris, 1971).

(2) *Control systems can create new experts, giving them power and autonomy.* When a new system is implemented, one group will gain power (i.e. more control over information). At the same time, another group will have its power reduced and will likely resist the control system (Pettigrew, 1972).

(3) *Control systems have the potential to measure individual performance more accurately and completely.* Some control systems are resisted because they measure aspects of performance that have not previously been measured. Those employees who feel that increased accuracy in performance data will reflect positively on their performance, will accept the new system. Many, however, may see the more objective evaluation system as threatening to their job security, status and power in the organization and as a result will resist the system.

(4) *Control systems can change the social structure of an organization.* Controls such as pay incentive plans, work measurement systems and computerized management information systems can affect the social relationships among workers (e.g. decrease social contact among workers). Zuboff (1982) points out that such systems may interfere with the informal workplace community and decrease commitment to the job and to the organization, even when tasks are tedious and uninteresting.

(5) *Control systems can reduce opportunities for intrinsic need satisfaction.* The work of Hackman and Lawler (1971) indicates that a task must have the

following four characteristics if a person is to experience psychological success and intrinsic satisfaction from good job performance:

(a) a high degree of task identity,
(b) feedback available about performance,
(c) a high degree of autonomy possible in how task is performed,
(d) a high degree of variety.

Control systems can, for example, reduce the amount of autonomy by specifying exactly how things have to be done in a job, and prevent the person from feeling intrinsic satisfaction from the work.

Lawler clearly summarizes the situations in which resistance to control systems is most likely to be present (Lawler, 1976, p. 1274):

(1) the control system measures performance in a new area;
(2) the control system replaces a system that people have a high investment in maintaining;
(3) the standards are set without participation of workers;
(4) the results are not fed back to the people whose performance is being measured;
(5) the results are fed to higher levels in the organization and are used by the reward system;
(6) the people who are affected are relatively satisfied with things as they are, and see themselves as committed to the organization;
(7) the people affected by the system are low in self-esteem and authoritarianism.

Falsification of Control System Data

Researchers have shown that control systems produce two kinds of invalid data. The first type of invalid data concerns what *can* be done. This falsification of data often occurs in standard setting (Lawler, 1976), and in budget preparation (Argyris, 1964) where workers consciously make invalid estimates of what is possible. The second type of invalid data relates to what *has* been done. People may feed the system falsified data in order to cover up errors or poor performance, to discourage use of the system by others (e.g. discourage management from relying on it), or to provide data that are demanded by the system, even when the data are not and cannot be collected.

This type of dysfunctional behavior will most likely occur when the following conditions exist in the system:

(1) the data are subjective in nature;
(2) the data are measuring a dimension that the individual sees as reflecting on his or her competence in an important area;

(3) standards are set by a process that does not allow participation by the individuals being measured, and the standards are seen as being unreasonable;
(4) the individual has control over the information gathering and the discriminator function;
(5) the information is given to an individual's superior, who uses it to evaluate the individual for the giving and withholding of significant rewards and punishments;
(6) the individual values the rewards that are related to the data and the individual is alienated from the system;
(7) the activity is not one that produces clear-cut outcomes and is difficult to measure;
(8) the activity is not one that is so important to the functioning of the organization that if it is not performed adequately the organization will cease functioning (Lawler, 1976, p. 1264).

Potential Positive Effects of Control Systems

Although the empirical findings point to the dysfunctional consequences of control systems, many writers also discuss their potential positive impacts. A control system may have positive effects on performance if it is designed 'properly.'

(1) It can provide feedback to individuals about task performance. The importance of feedback as a motivator has been stressed by several researchers (Festinger, 1954; Pettigrew, 1972; Hackman and Lawler, 1971). When people receive information on how well they are performing, they tend to perform their tasks better (Vroom, 1964). It is also important for people to compare their performance with that of others (Festinger, 1954; Pettigrew, 1972). Most people exhibit 'motivational problems' when they are protected from data about how well they are performing (Hackman and Oldham, 1980).
(2) It can structure tasks and indicate how performance on tasks will be measured. Most people want their job defined to some extent (Lawler, 1976).
(3) It can provide the opportunity for self-control of behavior. As an example, an employee has some control over what rewards are received in organizations where pay incentives are in effect (McGregor, 1960).
(4) It can cause extrinsic motivation. Control systems that are tied into rewards often act as the external motivator for an individual (i.e. the drive to perform the job) (Kerr, 1982).

Related to the question of the design of a proper control system is the employees' perception of its fairness. The following have been identified as factors affecting employees' perceptions of fairness and accuracy of a performance evaluation system (Landy, Barnes and Murphy, 1978; Pavett, 1983):

(1) frequency of evaluation,
(2) identification of goals,
(3) supervisor's knowledge of subordinate's level of performance and job duties,
(4) clarification of supervisor's expectations of performance,
(5) personal confidence of the employee being evaluated.

An organization must ensure that it has employees who are both capable and *willing* to use the advanced technology to achieve organizational objectives (Steers and Porter, 1983). Some researchers agree that computers can be used to provide the individual with data that were previously blocked from him or her, and stress that such immediate feedback can act as a motivator (Hackman *et al.*, 1979). Direct feedback sources outside the supervisor's control can be considered very valuable by individuals (Greller, 1980; Pavett, 1983).

Although relevant feedback may be valuable, Likert (1977) points out that the introduction of work measurement procedures and work standards may be accompanied by long-term adverse effects. These could include less favorable attitudes (e.g. resentment), or group efforts to defeat the organization's production goal (e.g. by restriction of output). In this context, Likert refers to a study of the impacts of a work measurement program on productivity and on morale at IBM manufacturing plants (Likert, 1977). There is evidence that, in spite of short-term gains in productivity, the time lag that occurs between changes in financial variables and changes in human performance may cause serious problems for an organization in productivity assessment (Likert, 1977; Walton and Vittori, 1983).

BIBLIOGRAPHY

Ackoff, R. L. (1967) 'Management misinformation systems', *Management Science*, **14**(4), 147–156.
Argyris, C. (1951) *The Impact of Budgets on People*, Controllership Foundation, New York.
Argyris, C. (1964) *Integrating the Individual and the Organization*, John Wiley, New York.
Argyris, C. (1971) 'Management information systems: challenge to rationality and emotionality', *Management Science*, **17**(6), B275–B292.
Babbage, C. (1835) *On the Economy of Machinery and Manufacturers*, C. Knight, London.
Bair, J. H. (1982) 'Productivity assessment of office information systems technology', in R. Landau, J. Bair and J. Seigman (eds), *Emerging Office Systems*, Ablex, Norwood, NJ, pp. 159–186.
Blau, P. M. (1963) *The Dynamics of Bureaucracy*, University of Chicago Press, Chicago, IL.

Brayfield, A. H. and Crockett, W. H. (1955) 'Employee attitudes and employee performance', *Psychological Bulletin*, **52**(5), 396–424.

Brown, W. B. (1968) 'The organization and socio-technical controls', *MSU Business Topics*, Winter, pp. 39–46.

Chorafas, D. N. (1982) *Office Automation: The Productivity Challenge*, Prentice-Hall, Englewood Cliffs, NJ.

Cooley, M. (1980) 'Computerization—Taylor's latest disguise', *Economic and Industrial Democracy*, **1**, 523–539.

Cummings, L. L. and Schwab, D. P. (1973) *Performance in Organizations: Determinants and Appraisal*, Scott, Foresman, Glenview, IL.

Eason, K. D. (1982) 'The processing of introducing information technology', *Behaviour and Information Technology*, **1**(2), 197–213.

Eilon, S. (1962) 'Problems in studying management control', *International Journal of Production Research*, **1**, pp. 134–149.

Emery, F. E. (1969) *Organizational Planning and Control Systems*, Bell Laboratories internal memorandum.

Emery, F. E. and Trist, E. L. (1960) 'Socio-technical systems', in M. Verhulst (ed.), *Management Science, Models and Techniques*, Vol. 2, Pergamon, Oxford, pp. 83–97.

Fayol, H. (1949) *General and Industrial Administration*, Pitman, London.

Festinger, L. (1954) 'A theory of social comparison processes', *Human Relations*, **7**, 117–140.

Fiedler, F. E., Chemers, M. M. and Mahar, L. (1977) *Improving Leadership Effectiveness: The Leader Match Concept*, Wiley, New York.

Fitch, D. S. (1982) *Increasing Productivity in the Microcomputer Age*, Addison-Wesley, Reading, MA.

Gotlieb, C. C. and Borodin, A. (1973) *Social Issues in Computing*, Academic Press, New York.

Granick, D. (1960) *The Red Executive: a study of the organization man in Russian industry*, Doubleday, Garden City, New York.

Gregory, J. and Nussbaum, K. (1982) 'Race against time: automation in the office', *Office: Technology and People*, **1**, 197–236.

Greller, M. M. (1980) 'Evaluations of feedback sources as a function of role and organizational level', *Journal of Applied Psychology*, **65**, 24–27.

Griffin, R. W. (1982) 'Perceived task characteristics and employee productivity and satisfaction', *Human Relations*, **35**(10), 927–938.

Griffin, R. W., Welsh, A. and Moorhead, G. (1981) 'Perceived task characteristics and employee performance: a literature review', *Academy of Management Review*, **6**(4), 655–664.

Hackman, J. R. and Lawler, E. E. III (1971) 'Employee reactions to job characteristics', *Journal of Applied Psychology Monograph*.

Hackman, J. R. and Oldham, G. R. (1980) *Work Redesign*, Addison-Wesley, Reading, MA.

Hackman, J. R., Oldham, G. R., Janson, R. and Purdy, K. (1979) 'A new strategy for job enrichment', in R. B. Peterson and L. Tracy (eds), *Readings in Systematic Management of Human Resources*, Addison Wesley, Reading, MA, pp. 81–102.

Henderson, R. (1980) *Performance Appraisal: Theory to Practice*, Reston, Va, Reston.

Herzberg, F. (1968) 'One more time: how do you motivate employees', *Harvard Business Review*, January–February, pp. 46, 54–62.

Hill, S. (1981) *Competition and Control at Work*, MIT Press, Cambridge, MA, pp. 103–107.

Howell, W. C. (1982) 'An overview of models, methods and problems', in W. C. Howell and E. A. Fleishman (eds), *Human Performance and Productivity*, Vol. 2, Lawrence Erlbaum, Hillsdale, NJ, pp. 1–29.

Johnson, B. M., Taylor, J. C., Smith, D. R. and Cline, T. R. (1983) 'Innovation in word processing', National Science Foundation Project ISI811079, National Science Foundation.

Kasurak, P. C., Tan, C. and Wolchuk, R. (1982) 'Management and the human resource impact of the electronic office', *Optimum*, Supply and Services Canada, Vol. 13–1, pp. 57–68.

Kelly, H. H. and Thibaut, J. W. (1969) 'Group problem solving', in G. Lindzey and E. Aronson (eds), *The Handbook of Social Psychology*, Vol. IV, 2nd edn, Addison-Wesley, Toronto, Canada, pp. 1–101.

Kerr, S. (1973) 'Some modifications in MBO as an OD strategy', *Academy of Management Proceedings*, pp. 39–42.

Kerr, S. (1982) 'On the folly of rewarding "A", while hoping for "B" ', in D. Hampton (ed.), *Organizational Behavior and the Practice of Management*, Scott, Foresman & Co., Glenview, IL.

Klein, L. (1976) *New Forms of Work Organization*, Cambridge University Press, Cambridge.

Kling, R. (1973) 'Towards a person-centered computer technology', *Proceedings of the ACM National Conference*, pp. 387–391.

Kling, R. and Scacchi, W. (1980) 'Computing as social action: the social dynamics of computing in complex organizations', *Advances in Computers*, **19**, 249–327.

Kling, R. and Scacchi, W. (1982) 'The web of computing: computer technology as social organization', *Advances in Computers*, **21**, 1–90.

Landy, F. J. and Farr, J. L. (1980) 'Performance rating', *Psychological Bulletin*, **87**(1), 72–107.

Landy, F. J. and Trumbo, D. A. (1980) *Psychology of Work Behavior*, Dorsey Press, Homewood, IL.

Landy, F. J., Barnes, J. and Murphy, K. (1978) 'Correlates of perceived fairness and accuracy of performance evaluation', *Journal of Applied Psychology*, **63**(6), 751–754.

Latham, G. P. and Wexley, K. N. (1981) *Increasing Productivity through Performance Appraisal*, Addison-Wesley, Reading MA.

Lawler, E. E. III (1976) 'Control systems in organizations', in M. D. Dunnette (ed.) *Handbook of Industrial and Organizational Psychology*, Rand McNally, Chicago, Il, pp. 1247–1291.

Lawler, E. E. III and Rhode, J. G. (1976) 'Dysfunctional effects of control systems', in E. E. Lawler and J. G. Rhode (eds), *Information and Control in Organizations*, Goodyear Publishing, Palisades, CA, pp. 84–94.

Lehrer, R. N. (1983) *White Collar Productivity*, McGraw-Hill, New York.

Levinson, H. (1976) 'Appraisal of what performance?', *Harvard Business Review*, July–August, p. 30.

Lewin, K. (1938) *The Conceptual Representation and the Measurement of Psychological Forces*, Duke Univeristy Press, Durham, NC.

Licker, P. S. (1982) 'The effects of the automated office on professional performance', Working Paper No. 16–82, Faculty of Management, University of Calgary, July.

Likert, R. (1977) 'Management styles and the human component', *Management Review*, October, pp. 23–28, 43–45.

Markus, M. L. (1983) 'Power, politics and MIS implementation', *Communications of the ACM*, **26**(6), 430–444.

Maslow, A. H. (1954) *Motivation and Personality*, Harper and Row, New York.

Mayo, E. (1933) 'Hawthorne and the Western Electric Company', Chapter 4 in *The Human Problems of an Industrial Civilization*, pp. 60–76, Macmillan, New York.

McGregor, D. (1960) *The Human Side of Enterprise*, McGraw Hill, New York.

Meltzer, H. and Nord, W. (eds), (1981) *Making Organizations Humane and Productive*, Wiley, New York.

Merton, R. K. (1940) 'Bureaucratic structure and personality', *Social Forces*, pp. 560–568.

Mintzberg, H. (1979) *The Structuring of Organizations*, Prentice-Hall, Englewood Cliffs, NJ.

Morrisey, G. L. (1983) *Performance Appraisals for Business and Industry*, Addison-Wesley, Reading, MA.

Myers, C. (ed.) (1967) *The Impact of Computers on Management*, MIT Press, Cambridge, MA.

Nolan, R. E. (1983) 'Work measurement', in R. Lehrer (ed.), *White Collar Productivity*, McGraw Hill, New York, pp. 111–157.

Olson, M. and Lucas, H. C. Jr (1982) 'The impact of office automation on the organization: some implications for research and practice', *Communications of the ACM*, **25**(11), 838–847.

Pavett, C. M. (1983) 'Evaluation of the impact of feedback on performance and motivation', *Human Relations*, **36**(7), 641–654.

Pettigrew, W. M. (1972) 'Information control as a power resource', *Sociology*, **6**, 188–204.

Reeves, T. and Woodward, J. (1970) 'The study of managerial controls', in J. Woodward (ed.), *Industrial Organization: Behaviour and Control*, Oxford, London, pp. 42–73.

Rice, A. K. (1958) 'Productivity and social organization: the Ahmedabod experiment', Tavistock Institute, London.

Ridgway, V. F. (1956) 'Dysfunctional consequences of performance measurements', *Administrative Science Quarterly*, **2**(2), 240–247.

Simon, H. (1948) *Administrative Behaviour*, Macmillan, New York.

Skinner, B. F. (1953) *Science and Human Behavior*, Free Press, New York.

Stabell, C. B. (1982) 'Office productivity: a micro-economic framework for empirical research', *Office: Technology and People*, **1**, 91–106.

Steers, R. and Porter, L. (1983) *Motivation and Work Behaviour*, McGraw Hill, New York.

Stewart, R. (1971) *How Computers Affect Management*, Macmillan, London.

Taylor, F. (1947) *Scientific Management*, Harper & Row, New York, pp. 1–101.

Thompson, P. H. and Dalton, G. W. (1970) 'Performance appraisal: managers beware', *Harvard Business Review*, January–February, pp. 149–157.

Trist, E. L. and Bamforth, K. W. (1951) 'Some social and psychological consequences of the longwall method of coal-getting', *Human Relations*, **4**, 3–38.

Vroom, V. H. (1964) *Work and motivation*, Wiley, New York.

Walton, R. E. and Vittori, W. (1983) 'New information technology: organizational problem or opportunity?', *Office: Technology and People*, **1**, 249–273.

Whisler, T. L. (1970) *The Impact of Computers on Organizations*, Praeger, New York, pp. 94–124, 142–180.

Woodward, J. (1970) *Industrial Organization: Behaviour and Control*, Oxford University Press, London.

Yager, E. (1981) 'A critique of performance appraisal systems', *Personnel Journal*, **60**(2), 131.

Zuboff, S. (1982) 'Computer mediated work—the emerging managerial challenge', *Office: Technology and People*, **1**, 237–243.

AUTHOR INDEX

SUBJECT INDEX